W9-CBB-239

The Dolciani Mathematical Expositions

NUMBER TWENTY-THREE

The Beginnings and Evolution of Algebra

I. G. Bashmakova
and
G. S. Smirnova

Translated from the Russian by Abe Shenitzer
with the editorial assistance of David A. Cox

Published and distributed by
The Mathematical Association of America

THE
DOLCIANI MATHEMATICAL EXPOSITIONS

Published by
THE MATHEMATICAL ASSOCIATION OF AMERICA

———

The Beginnings and Evolution
of Algebra

©2000 by
The Mathematical Association of America (Incorporated)
Library of Congress Catalog Card Number 99-68950

Complete Set ISBN 0-88385-300-0
Vol. 19 ISBN 0-88385-329-9

Printed in the United States of America

Current printing (last digit):
10 9 8 7 6 5 4 3 2 1

The DOLCIANI MATHEMATICAL EXPOSITIONS series of the Mathematical Association of America was established through a generous gift to the Association from Mary P. Dolciani, Professor of Mathematics at Hunter College of the City University of New York. In making the gift, Professor Dolciani, herself an exceptionally talented and successful expositor of mathematics, had the purpose of furthering the ideal of excellence in mathematical exposition.

The Association, for its part, was delighted to accept the gracious gesture initiating the revolving fund for this series from one who has served the Association with distinction, both as a member of the Committee on Publications and as a member of the Board of Governors. It was with genuine pleasure that the Board chose to name the series in her honor.

The books in the series are selected for their lucid expository style and stimulating mathematical content. Typically, they contain an ample supply of exercises, many with accompanying solutions. They are intended to be sufficiently elementary for the undergraduate and even the mathematically inclined high-school student to understand and enjoy, but also to be interesting and sometimes challenging to the more advanced mathematician.

MAA Service Center
P. O. Box 91112
Washington, DC 20090-1112
1-800-331-1622   fax: 301-206-9789

Notes by translator and editor

Translator's note. I wish to thank the following people: David A. Cox for his exceptional helpfulness and for the enlightening comments which he added in a number of places in the translation; Ed Barbeau, Bob Burckel, Giuliana Davidoff, and Dan Velleman, members of the *Dolciani Mathematical Expositions Board,* for their comments and constructive criticism; John Stillwell for eliminating a number of errors; Sarah Shenitzer for spotting errors and linguistic rough spots in an "almost final" version of the translation; and, last but not least, Don Albers, Elaine Pedreira, and Beverly Ruedi for managing the whole enterprise without a hitch.

My special thanks go to Dan Velleman. In addition to correcting various errors Dan suggested a number of helpful modifications of key statements.

Note by translator and editor. This is a history of algebra that starts with an account of its Babylonian beginnings and, apart from sketchy remarks about developments in the 20th century, ends with a substantial account of major developments in the 19th century.

This is not a textbook. While the early parts of the book can be read by people with no more than a high school knowledge of mathematics, the later parts of the book require of the reader familiarity with the material usually presented in a course in abstract algebra covering groups, rings, fields, and Galois theory. Readers with such background will be able to appreciate the full story of polynomial equations in one variable as well as the story of Diophantine analysis that is nowadays at the very center of mathematical research.

Contents

INTRODUCTION

The 19th century witnessed fundamental transformations of the major divisions of mathematics—geometry, algebra, and mathematical analysis. Of these, the qualitative changes in geometry, especially the creation of non-Euclidean and multidimensional geometries, may well have had the profoundest effect on the mathematical imagination. As a result—as noted by Bourbaki—classical geometry became a universal language for the interpretation of mathematical facts and theories. One could talk of a geometric style of thinking (A. N. Kolmogorov, *The Great Soviet Encyclopedia*, 2nd ed., Vol. 26 (Russian)).

But at the end of the 20th century it became clear that one would be equally justified in calling the 19th century the age of new algebra, and in talking of the algebraization of mathematics and the elaboration of an algebraic style of thinking. Indeed, until the 19th century algebra was largely the science of (determinate and indeterminate) equations, whereas in the 19th century there appeared in it completely new concepts and objects, such as groups, rings, fields, ideals, matrices, algebras, and many others. Their study resulted in the development of new methods and conceptions, and this brought about a changed view of the subject matter of algebra. Specifically, the task of algebra was now seen to be the study of systems of arbitrary nature "for which there are defined operations with properties more or less similar to those of addition and multiplication of numbers" (A. G. Kurosh and O. Yu. Schmidt, *The Great Soviet Encyclopedia*, 2nd ed., Vol. 1 (Russian)). These operations were called composition laws and their basic properties were given by systems of axioms.

The methods of this new, so-called "modern", algebra quickly entered other areas of mathematics. The 19th century witnessed the construction of algebraic number theory, the development of the first stages of algebraic geometry, and the rigorization of the theory of Riemann surfaces by algebraic means. But what was startling was the "victorious march" of group theory, which is today an indispensable ingredient of every area of mathematics. We will talk about this in detail in Chapter VII (5).

We add that functional analysis, which came into being at the end of the 19th century, was constructed as linear algebra of infinite-dimensional spaces.

Finally, from the first quarter of the 20th century onward, algebraic methods were intensively applied in physics and brought about its radical transformation. There came into being matrix mechanics, the theory of spinors, and the subject of symmetry, which plays so important a role today. Already in the 1890s E. S. Fedorov applied the theory of finite groups to crystallography, and in this way managed to solve the problem of classification of regular point systems in space (we recall that there are 230 Fedorov space groups and that their classification could not be carried out without the use of group theory). Similarly, more general groups and their representations are used to classify elementary particles and their motions.

How did algebra arise? What are its subject matter and methods? How have they changed in the process of its evolution? These are the questions we will try to answer in the present essay. Before we do so we note that the view of algebra as the science of operations defined on sets of arbitrary objects came into being quite late, probably only in the 1930s.

In its evolution algebra passed through different phases during which it was thought of differently. Views of its subject matter, methods, and aims changed. There is hardly a branch of mathematics whose evolution involved as many surprising metamorphoses as that of algebra. Nevertheless, if we cast a retrospective glance at its development, then we see that the characteristic feature of algebra from its very first steps and practically until the appearance (at the beginning of the 19th century) of noncommutative and nonassociative systems was the study of laws of composition and of their fundamental properties: commutativity of addition and multiplication, distributivity of multiplication over addition, rules for multiplication of binomials and for raising them to powers, rules for operating with equations, and so on. This being so, we will begin our study of the history of algebra from the time of the discovery and application of the simplest laws of composition.

When characterizing each of the fundamental stages of the evolution of algebra we will focus our attention on the problems that faced it and stimulated its development as well as on the basic ideas and methods used in their investigation.

Modern history of mathematics seems to be dominated by the view that up to the 1830s the mainspring of the development of algebra was the investigation and solution of determinate algebraic equations, and especially their solution by radicals. We will show that this viewpoint is one-sided and gives a distorted representation of its evolution. In short, we claim that the role of indeterminate equations in the development of algebra was no less important than that of determinate equations.[1]

We note also that the rate and phases of the evolution of algebra do not always correspond to the rate and periods of evolution of mathematics as a whole. In our account the history of algebra is divided into the following basic stages.

1. Numerical algebra of ancient Babylonia (a phase that coincides in time with the first period of the history of mathematics, i.e., the period of accumulation of mathematical knowledge).

2. Geometric algebra of classical antiquity (a phase, lasting from the 5th to the 1st century BCE, which corresponds to the first half of the second period of the history of mathematics, the period of the transformation of mathematics into an abstract theoretical science).

3. The rise of literal algebra (from its birth to the creation of literal calculus; this phase began in the 1st century CE and lasted until the end of the 16th century, i.e., it began in the second half of the second period of the history of mathematics and lasted until the end of its third period, the period of development of elementary mathematics).

4. Creation of the theory of algebraic equations (a phase that comprises the development of algebra in the 17th and 18th centuries and ends in the 1830s).

5. Formation of the foundations of modern algebra (a phase lasting from the 1830s to the 1930s).

We will provide detailed characterizations of these phases in parallel with the exposition of the relevant material. The last phase of the evolution of algebra, the one that began some fifty odd years ago, cannot as yet be classified as a component of history.

Editor's notes

[1] A system of polynomial equations is said to be indeterminate if it has fewer equations than variables, and to be determinate if the number of equations

equals or exceeds the number of variables. Thus

$$x + y = a, \quad a \text{ constant,}$$

$$xy = 1,$$

is a determinate system, while

$$x^2 + y^2 = z^2$$

is indeterminate. Note that polynomial equations in one variable, such as quadratic or cubic equations, are determinate.

The term "indeterminate" is most commonly applied to systems with fewer equations than variables for which integer or rational solutions are sought. Such systems, also called "Diophantine", have been particularly influential in the development of algebra (see Chapter VIII).

CHAPTER 1

Elements of algebra in ancient Babylonia

1. Babylonian numerical algebra

The elements of algebra were created in ancient Babylonia, two millennia BCE. Modern scholars have found out about this only in the 1930s. In the middle of the 19th century a large number (by now over 500,000) of clay tablets with cuneiform writing were found in Mesopotamia. Some of them were mathematical, but their contents were first deciphered only between 1929 and 1930 and disclosed a new world of ancient Babylonian mathematics. Major credit for these discoveries goes to O. Neugebauer, whose work created the foundation for a multitude of subsequent investigations. The names of some of the important scholars in this area are: F. Thureau-Dangin, A. Sachs, M. Ya. Vygodskiĭ, I. N. Veselovskiĭ, A. A. Vaĭman, and E. M. Bruins.

It turned out that the mathematical tablets were either "table texts" or "problem texts". The table texts contained multiplication tables, tables of squares, of cubes, and so on. The problem texts contained statements and solutions of problems. The tablets belong to two clearly delimited periods separated by a long intermediate period. Most of the tablets are "old Babylonian", i.e., they date back to the Hammurapi dynasty (1800–1600 BCE). The rest were made during the Seleucid period (from the 3rd to the 2nd century BCE), i.e., in the Hellenistic period. During the intermediate period the language and the manner of writing the symbols changed, and these changes made possible the precise dating of the tablets. What did not change was the mathematical content of the texts. No significant development took place in all these centuries.

The most surprising fact is that most of the problem texts contained problems of the same type that reduced to the solution of quadratic equations. Thus close to 2000 years BCE the Babylonians could solve quadratic equations

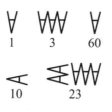

and carry out algebraic transformations! This remarkable discovery pushed back the beginnings of algebra by 13 centuries—from the earlier estimate of the 5th century BCE to the 18th century BCE!

Before considering the problems we will say a few words about the table texts. The Babylonians used just two symbols (two digits!), a horizontal wedge and a vertical wedge, for denoting all numbers (Figure 1). These symbols were impressed with a thin stick on clay tablets that were later baked. In spite of the fact that the Babylonian numeral system employed just two digits, its base was 60. The numbers from 1 to 59 were recorded using the additive principle, with a vertical wedge for 1 and a horizontal wedge for 10. The symbol for 23 is shown in Figure 1.

The number 60 was also denoted by a single vertical wedge. Thus ⱽⱽⱽ could be read as 83. We said "could" advisedly, because the same group of symbols could also be interpreted as $1\frac{23}{60}$, or as $60^2 + 23$, or as any of the numbers of the form $60^k + 23 \cdot 60^{k_1}$, where $k_1 < k$. Of course, this nonuniqueness was due to a lack of an analogue of our zero. It should be noted that during the Seleucid period a separating symbol, a prototype of our zero, was used to indicate a missing sexagesimal order, but this symbol was never used at the end of a number.

Briefly, the Babylonian numeral system was *positional, nonunique*, and had a base of 60.

What made up, to some extent, for the awkwardness due to the lack of zero was that the Babylonians could record integers as well as sexagesimal fractions in a uniform way, i.e., they had a positional notation as well as a systematic method for writing fractions.

The large base of 60 called for a huge multiplication table that went up to 59×59. Luckily, children were not required to memorize it. The large number of dug up multiplication tables (they represent the largest fraction of all unearthed table texts) of this kind shows that they were, so to say, mass produced for the use of pupils.

Other tables dealt with division. To divide M by N the Babylonians multiplied M by $N' = 1/N$. Hence the large number of tables of multiplicative inverses that were also used by pupils.

Finally, there were tables of squares and cubes of natural numbers, tables of sums of squares and cubes, and tables of square roots (expressed in sexagesimal fractions). In connection with square roots we note the Babylonians' use of a very convenient formula: if $\sqrt{a} \approx x_1$, then they took as the next approximation $x_2 = \frac{1}{2}(x_1 + \frac{a}{x_1})$. This (rapidly converging) process could be iterated.

Traces of the Babylonian numeral system have survived to this day: we divide an hour into 60 minutes and a minute into 60 seconds. We do the same when dividing a degree into minutes and seconds. The source of this tradition is astronomy. The Babylonians were the first to carry out systematic observations of the starry sky. They constructed a calendar, computed the periods of the moon and the planets, and could predict eclipses of the moon and the sun. Their astronomical knowledge was inherited by the Greeks, who adopted their astronomical tables together with their sexagesimal numeral system. The great astronomer and mathematician Claudius Ptolemy (2nd century CE) used the sexagesimal system in his fundamental work known under the Arabized name of *Almagest*. The *Almagest* was the basic astronomical work in the medieval East as well as in Europe, where it was still in use in the 16th and 17th centuries.

Next we turn to the problem texts.

Using modern notation, we could state the problems in the cuneiform tablets as follows (a, b, and c are given and x and y are to be found):

$$\begin{cases} x \pm y = a, \\ xy = b; \end{cases} \tag{1}$$

$$\begin{cases} x \pm y = a, \\ x^2 + y^2 = b; \end{cases} \tag{2}$$

$$ax^2 \pm bx = c. \tag{3}$$

The Babylonians denoted unknowns by special terms (for which they used single-mark Sumerian words, whereas the text itself was in Akkadian) rather than by symbols: "length" (for our x), "width" (y) (invariably $x > y$), and "area" (xy). Sometimes there was mention of "sides of squares". If necessary, a third unknown, "depth" (z), was introduced, and the product xyz was called "volume". While the terminology was explicitly geometric, all unknowns were regarded as numbers, so that the Babylonians freely added areas to lengths or widths, and so on. There were also more abstract statements of problems: there is a series of problems that involve the computation of a "factor" and its "inverse", i.e., of two numbers whose product is 1.

We call problems (1)–(3) canonical. They were solved by an algorithm that fully corresponds to our formula for the solution of a quadratic equation. We state one such problem: "A factor and its inverse 2; 30." In our notation, the problem can be stated as follows:

$$\begin{cases} x + y = a, \\ xy = 1. \end{cases} \tag{4}$$

Here x is the "factor", y is the "inverse", and $a = 2; 30$. In decimal notation 2; 30 is 2.5 (2; 30 stands for $2\frac{30}{60}$, i.e., for 2.5).

What follows is a rhetorical solution of the problem accompanied by two "translations", one decimal and the other literal.

Solution

"factor and inverse 2; 30"	$x + y = 2.5,$ $xy = 1$	$x + y = a,$ $xy = b$
1. "by 0;30 multiply: 1;15"	$2.5 \cdot \frac{1}{2} = 1.25$	$\frac{a}{2}$
2. "1;15 by 1;15 multiply: 1;33,45"	$1.25 \cdot 1.25 = 1.5625$	$\frac{a}{2} \cdot \frac{a}{2} = \left(\frac{a}{2}\right)^2$
3. "1 subtract from this: 0;33,45"	$1.5625 - 1 = 0.5625$	$\left(\frac{a}{2}\right)^2 - b$
4. "what must one multiply by what to obtain 0;33,45? 0;45 "	$\sqrt{0.5625} = 0.75$	$\sqrt{\left(\frac{a}{2}\right)^2 - b}$
5. "0;45 to 1;15 add: 2-factor"	$1.25 + 0.75 = 2[= x]$	$\frac{a}{2} + \sqrt{\left(\frac{a}{2}\right)^2 - b}$ $= x$
6. "0;45 from 1;15 subtract: 0;30-inverse"	$1.25 - 0.75 = 2[= y]$	$\frac{a}{2} - \sqrt{\left(\frac{a}{2}\right)^2 - b}$ $= y$

We see that the solution recipe consists of the sequence of operations required for the solution of the quadratic equation $z^2 - az + b = 0$, which is equivalent to the system (4). The text contains no explanations. The generality of the algorithm was illustrated by applying it to a large number of problems of the same type. The text is obviously of the textbook variety.

How could this algorithm have been obtained? One probably used algebraic transformations to reduce (4) to the form $z^2 - B = 0$. If the solution $x = y = a/2$ failed to satisfy the conditions of the problem, then one put $x = a/2 + t$, $y = a/2 - t$, which meant that $xy = (a/2)^2 - t^2 = b$, i.e.,

$t^2 = (a/2)^2 - b$. What is striking is that in his *Arithmetic* (Book 1, Problem 27) Diophantus solved this system in just this way. Apparently, the Babylonian method survived for more than two millennia!

To carry out the transformation just described the Babylonian calculator had to know: 1) the substitution rule; and 2) the rule for removing parentheses, which relied on the distributivity of multiplication over addition,

$$a(b + c) = ab + ac. \tag{5}$$

The latter was the basis for the formula

$$(a + b)(a - b) = a^2 - b^2. \tag{6}$$

Can we credit the Babylonians of this distant era with such knowledge? We are about to show that we have very good reasons to do so.

To this end we consider problems of "noncanonical" nature, i.e., problems that do not directly reduce to the equations (1)–(3). We give two examples from O. Neugebauer's work (O. Neugebauer, *Mathematische Keilschrifttexte.* Quellen und Studien zur Geschichte der Mathematik, Astronomie und Physik. Abt. A. Berlin, 1930–1937).

A problem from text AO 8862 from Senkere (from the period of the Hammurapi dynasty) states: "Length, width. Length and width I multiplied and obtained area. Then I added the excess of length over width to the area: I got 3,3. Then I added length and width: 27. One asks: length, width, and area" (Vol. I, p. 133).

Using modern notation we rewrite the problem as follows:

$$\begin{cases} xy + (x - y) = a, \\ x + y = b, \end{cases} \tag{7}$$

where $a = 183$ and $b = 27$. The solution begins with the words: "You will do thus:

$$27 + 3,3 = 3,30,$$

$$2 + 27 = 29".$$

(Here 3,3 stands for $3\cdot60+3$ and 3,30 stands for $3\cdot60+30$.) This is followed by the solution of the system

$$\begin{cases} uv = 3,30, \\ u + v = 29. \end{cases} \tag{8}$$

What is the connection between the systems (7) and (8)? The Babylonian calculator finds

$$x = u = 15; \qquad y = v - 2 = 14 - 2 = 12.$$

If we analyze this solution, then we arrive at the conclusion that the first step was the addition of the left sides of the given equations:

$$xy + (x - y) + (x + y) = xy + 2x.$$

Next the "length" x, the common factor on the right, was factored out:

$$xy + 2x = x(y + 2).$$

This was followed by the substitution

$$v = y + 2.$$

Now $x + v = x + y + 2 = 29$. We are down to the system (8).

Thus it is safe to say that the Babylonians knew the substitution rule as well as the rule for common factors!

The following is another noncanonical problem:

"I added the area of my two squares: 25,25. (The side) of the second square equals 2/3 of the side of the first and another 5 GAR [="add"]" (ibidem, Vol. III, p. 8, no. 14). It is equivalent to the system

$$\begin{cases} x^2 + y^2 = a \\ y = kx + b, \end{cases} \tag{9}$$

where a=25,25 (1525 in the decimal system), $k = 2/3$, and $b = 5$.

The text begins with the computation of the following three coefficients:

$$p = 1 + (2/3)^2 = 1 + (0; 40)^2 = 1; 26, 40,$$

$$q = 5 \cdot 0; 40 = 3; 20, \tag{10}$$

$$r = 25, 25 - 5^2 = 25, 0.$$

Then the solution is found by using the formula

$$x = p^{-1} \left(\sqrt{pr + q^2} - q \right), \qquad y = \frac{2}{3}x + 5.$$

This means that in order to find the length x one had to solve the quadratic equation

$$px^2 + 2qx = r, \tag{11}$$

where p, q, and r are determined in (10). In order to obtain equation (11) the Babylonian calculator had to: square $y = 0; 40x + 5$, i.e., use the formula

$$(a + b)^2 = a^2 + b^2 + 2ab, \tag{12}$$

make a substitution, and reduce similar terms.

Thus two millennia BCE the Babylonians knew some laws of algebraic operations, made substitutions, and solved by algebraic methods quadratic equations and systems equivalent to quadratic equations. This justifies the claim that they knew the elements of algebra.

This algebra can be characterized as numerical, for it made no use of symbols. Formulas (6) and (12) were used to solve problems but were not considered in general form. We know of no proofs of these formulas. Many insights were probably the result of working out many numerical examples and of the application of incomplete induction. In the sequel we will discuss the origin of problems (1)–(3). Now we turn to problems of a kind that required the development of algebraic methods.

2. Indeterminate equations

Indeterminate equations were already considered during the rule of the Hammurapi dynasty. Many texts dealt with the problem of finding "Pythagorean triples", i.e., rational triples x, y, z satisfying the equation

$$x^2 + y^2 = z^2. \tag{13}$$

While the Babylonians knew many solutions of equation (13)—such as the triples $(3, 4, 5)$, $(5, 12, 13)$, $(8, 15, 17)$, and many others—it is not quite clear whether they knew the general formulas for its solutions. In this connection it should be mentioned that there is a text with a table of rational Pythagorean triples whose law of formation is controversial. Another equation considered side by side with equation (13) is the equation

$$u^2 + v^2 = 2w^2. \tag{14}$$

Both equations have a geometric origin. The first expresses the connection between the hypotenuse and legs of a right triangle. The second arose in connection with the problem, frequently encountered in Babylonian texts, of bisecting a given trapezoid (Figure 2) by means of a straight line parallel to its base. If we denote the upper base of the trapezoid by u, the lower base by v, and the dividing line by w, then it is easy to see that u, v, and w are

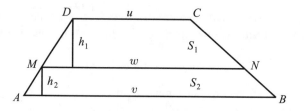

FIGURE 2

connected by equation (14). (Indeed, denote the area of the trapezoid by S, and the areas of its upper and lower halves by S_1 and S_2 respectively. Let h_1 be the altitude of the upper half and h_2 of the lower half. The condition $S_1 = S_2$ implies the equality $h_1/h_2 = (w + v)/(w + u)$, the condition $S_1 + S_2 = S$, the equality $h_1/h_2 = (u - w)/(w - v)$, and these two equalities imply equation (14).)

The Babylonians knew that there was a connection between equations (13) and (14): if x, y, z is a solution of (13), then $u = x - y, v = x + y, w = z$ is a solution of (14). Hence the table:

y	x	z	u	v
3	4	5	1	7
5	12	13	7	17
7	24	25	17	31

The number triples u, v, w are called Babylonian triples.

There are tables which indicate that the Babylonians could find infinitely many solutions of equation (14)! In fact, E. M. Bruins, who edited the Sousa texts, pointed out that the following procedure was used in the Sousa text no. 20: starting with a solution of (14) such as $u = 0; 15$, $v = 1; 45$, $w = 1; 15$ (obtained from the Babylonian triple $(1, 7, 5)$ by multiplication by $0; 15$, i.e., by $1/4$), one obtained a new solution in this way:

$$v_1 = 0; 48 \cdot 1; 45 + 0; 36 \cdot 0; 15 = 1; 33;$$

$$u_1 = 0; 36 \cdot 1; 45 - 0; 48 \cdot 0; 15 = 0; 51.$$

Similarly, one can obtain solutions (u_2, v_2, w), (u_3, v_3, w), and so on.

What is behind this procedure? The numbers by which one multiplies u and v to obtain u_1 and v_1 are $0; 48 = 48/60 = 4/5$ and $0; 36 = 3/5$ respectively. They are obtained from the right triangle $(3, 4, 5)$ by dividing its sides by 5. Now $(3/5)^2 + (4/5)^2 = 1$. This suggests that the Babylonians used the formula for composition of forms:

$$(u^2 + v^2)(\alpha^2 + \beta^2) = (\alpha u - \beta v)^2 + (\alpha v + \beta u)^2$$
$$= (\alpha u + \beta v)^2 + (\alpha v - \beta u)^2. \tag{15}$$

Both of these ways of representing $(u^2 + v^2)(\alpha^2 + \beta^2)$ as a sum of two squares are found in Babylonian texts. This justifies the conclusion that two millennia BCE the Babylonians knew the highly nontrivial formula (15). As we will see, this formula played an essential role in the work of Diophantus, in the mathematics of the medieval East, and in European mathematics between the 13th and 17th centuries.

In order to obtain successive solutions (u_k, v_k, w) of (14), the Babylonians used α and β such that $\alpha^2 + \beta^2 = 1$. Such values could be obtained from Pythagorean triples x, y, z by dividing all entries by z.

To sum up. Already during the first stage of its evolution algebra was influenced by the problem of solution of determinate (quadratic) equations as well as by the investigation and solution of indeterminate equations, and the most sophisticated Babylonian formula was associated with the solution of equations of the latter type.

3. The origin of the first algebraic problems

We mentioned earlier that geometry was the origin of problems that reduced to indeterminate equations. The origin of problems that reduced to quadratic equations is less clear. Land measurement gave rise to the problem: Given the length x and width y of a plot of land, find its area xy, or its area and perimeter $2(x + y)$. But it is clear that the problem of finding the sides of a plot of land given its area and perimeter has no practical value. How then did this problem arise? One can venture a guess that having solved the "direct" problem of determination of the area of a plot of land given its sides, the Babylonians tried to check the correctness of the solution and in this way arrived at the "inverse" problem of determining the sides of a plot of land given its area and perimeter.

The simplest direct problem is the following: Given the side a of a square find its area S. To solve it, we multiply the given number a by itself:

$$S = aa = a^2.$$

The inverse problem is to determine the side of a square given its area S. If x is the required side, then the inverse problem reduces to the "pure" quadratic equation

$$x^2 = S,$$

which is solved by extracting a square root, $x = \sqrt{S}$, an operation more complex than multiplication.

The next direct problem connected with land measurement is to find the area S of a plot of land given its sides a and b. This problem is also solved by multiplication: $S = ab$.

The inverse problem is to determine the sides x and y of a plot of land given its area S. This is an indeterminate problem. It can be made determinate in two ways:

1) by giving one of the sides, say $y = a$, in which case we obtain the equation

$$ax = S$$

which is solved by division $x = S/a$; or

2) by giving the sum (or difference) of the sides of the rectangle. In that case we obtain one of the canonical Babylonian systems:

$$xy = S, \qquad x + y = a.$$

The latter problem is basically more complex than the direct problem that gives rise to it. It was the analysis of this problem that led Babylonian scholars, close to four thousand years ago, to the discovery of the formula for the solution of quadratic equations. Gradually, such problems became an independent object of study and led to the rise and evolution of the elements of algebra.

Inverse problems played, and continue to play, a very basic role in the evolution of mathematics. We mention inverse trigonometric functions, inversion of series, the general notion of the inverse of a function and the determination of conditions for its existence and, finally, the inverse Galois (1811–1832) problem. All of these examples led to the consideration of new functions or to the creation of new theories. The first inverse problems are found in Babylonian mathematics, at the very sources of the evolution of algebra.

CHAPTER **2**

Ancient Greek "geometric algebra"

1. Transformation of mathematics into an abstract deductive science. Discovery of incommensurability

The second stage of the evolution of algebra coincides with the flourishing of classical Greek mathematics (from the 5th to the 2nd century BCE). At that time mathematical knowledge, accumulated through ages, was transformed into an abstract science based on a system of proofs—*mathematics*—and the first mathematical theories came into being.

Neither in ancient Babylonia nor in other pre-Greek civilizations do we find a single written proof. While scholars must have used individual arguments in support of their conclusions, such arguments were not at the center of their interests. What mattered was finding an effective recipe for the solution of a class of problems. The result could be exact or approximate (for example, when extracting roots or when computing certain areas or volumes). On the other hand, in Greece between the 6th and 5th centuries BCE interest was focussed on justification of propositions by proofs. Also, there was clear recognition that mathematical propositions deal with abstract objects: dimensionless points, lines with length but without width, and so on.

In the new, abstract, mathematics proofs performed a number of functions. The first of these functions was to *establish the truth* of a proposition. Proof came to be regarded as the only means of establishing truth. This is understandable: after all, all propositions of mathematics refer to abstract objects which can be realized only approximately in practice (take as an example the construction of an equilateral triangle). Also, most mathematical propositions refer to classes containing infinitely many such objects (for example, the class of all right triangles, of all isosceles triangles, of all primes, and so on). It is impossible to establish the truth of such propositions without proof.

Incidentally, it was realized early on that when constructing a theory one must start with a number of propositions—later known as axioms—that are accepted without proof. Aristotle pointed out that otherwise every proof would be infinite: proposition A would be proved on the basis of proposition B, proposition B on the basis of proposition C, and so on:

$$A \Leftarrow B \Leftarrow C \Leftarrow \cdots$$

The series of implications would extend to infinity and no proposition could ever be proved! Hence the need for selecting a finite number of propositions assumed to be true and for deducing all other propositions of the theory from them. Aristotle called a science constructed in this manner "demonstrative".

The second, no less important function of proof was to *establish connections among propositions*, to find out why certain propositions are true. Already Aristotle wrote in his *Posterior Analytics*: "To know what is and to know why it is are different types of knowledge" (*Analyt. post.* 78^a), and in his *Metaphysics* he wrote: "Those who have experience know the 'what' but not the 'why', whereas those who have mastered an art know the 'why', i.e., the cause" (*Meth.* 981^a). Faced with an unproved proposition we often do not know what theory it belongs to. Thus, in spite of their apparent resemblance, the theorems: "the sum of the angles in a triangle is equal to two right angles", and "an exterior angle of a triangle is greater than a nonadjacent interior angle" belong to different theories: the first to Euclidean geometry (its proof is fundamentally dependent on the parallel postulate) and the second to absolute geometry. It is only after the introduction of proofs that the propositions obtained from a single system of axioms begin to coalesce into a theory. In this way proofs determine a new mathematical structure.

The third function of proofs consists in the *discovery of new, previously unknown, propositions*. A striking instance of such a discovery is the proposition about the incommensurabiliy of the side and the diagonal of a square. We note that not only could this proposition not have been found intuitively but that it contradicts all accumulated human experience which tells us that there is a common measure for all magnitudes, i.e., that all magnitudes are commensurable. One had to command highly developed abstract thinking to prefer a result obtained by proof to experience and intuition.

We will not discuss the reasons for such a fundamental transformation of the system of mathematical knowledge; there are many papers and books devoted to this topic. We will only say that this transformation was undoubtedly connected with general qualitative changes in the cultural and political life of Greece. The question we *will* try to answer is where and when mathematics began to take the form of an abstract deductive science. This can be precisely determined as to time and place. In the first half of the 6th century BCE, the philosopher, astronomer, and merchant Thales of Miletus began to prove some propositions of geometry. Some of his proofs were "more general" and others "more empirical". And in the second half of the 5th century BCE the mathematician Hippocrates of Chios, who lived in Athens, wrote the first *Elements* of geometry, i.e., the first system based on proofs. This work has not come down to us; in fact, Euclid's famous work supplanted all *Elements* written before 300 BCE. What has come down to us is excerpts from Hippocrates' treatise on squarable lunes. These excerpts are written in what is still the standard style of mathematics. Hippocrates provides rigorous justifications of his assertions and never appeals to intuition by referring to a drawing. He knew not only Pythagoras' theorem but also the theorem about the side of a triangle subtended by an acute or obtuse angle as well as other theorems of plane geometry. This means that the transformation of mathematics and the creation of plane geometry took place some time between the middle of the 6th and the middle of the 5th centuries. But we know of the existence of just one school of mathematics at that time, namely the school of Pythagoras.

We also have direct testimony that points to Pythagoras as the man who transformed mathematics. In his commentaries on Book I of Euclid's *Elements*, the famous Neoplatonist philosopher Proclus had this to say: "Pythagoras ... transformed this science ($\varphi\iota\lambda o\sigma o\varphi\iota\alpha$) into a liberal form of education, examining its principles from the beginning and probing the theorems in an immmaterial and intellectual manner. He discovered the theory of proportionals and the construction of the cosmic figures" (Thomas 1939, p. 149).

We know that Pythagoras was born on the island of Samos, that in his youth he stayed in Egypt and studied with its priests, and that around 530 BCE he settled in Crotona, a Greek colony in Southern Italy (these colonies were known as Greater Greece), where he founded the brotherhood of Pythagoreans, which pursued scientific, religious-ethical, and political aims. Legends were told about Pythagoras already during his lifetime. By now it is difficult to separate truth from invention. The air of secrecy surrounding the brotherhood confused matters even more, so that we cannot tell which discoveries are due to Pythagoras and which to his students. This being so, we will talk about the mathematics of the early Pythagoreans.

The system of knowledge ($\mu\alpha\theta\eta\mu\alpha\tau\alpha$) of the early Pythagoreans consisted of four parts: arithmetic, geometry, the teaching of harmony, and astronomy. The Pythagoreans regarded arithmetic, i.e., the science of numbers, as the foundation of mathematics. By a number ($\alpha\rho\iota\theta\mu o\varsigma$) they meant a collection of units. They divided numbers into even and odd ones (much later, Plato described arithmetic as the study of even and odd) and proved the first theorem of divisibility theory: the product of two numbers is divisible by 2 if and only if at least one of them is divisible by 2. They also posed the problem of finding perfect numbers (i.e., numbers equal to the sum of their proper divisors). The Pythagoreans developed a theory of (positive) rational numbers. Since unity E was regarded as indivisible, it made no sense to talk about its "parts", i.e., fractions of the form $\frac{1}{n}E$ or $\frac{m}{n}E$. Instead, they talked about ratios of whole numbers $M : N$, $M = mE$, $N = nE$. Today we would say that they constructed the rational numbers as a theory of number pairs. This theory has come down to us through Euclid's *Elements* (Book VII). Euclid proves that the set of pairs of whole numbers with the same ratio,

$$M_1 : N_1 = M_2 : N_2 = M_3 : N_3 = \cdots,$$

contains a least pair P, Q, such that P and Q are relatively prime (i.e., the least pair corresponds to our irreducible fraction), and that the numbers M_k, N_k of any pair are equimultiples of P and Q, i.e., $M_k = rP$, $N_k = rQ$.

The early Pythagoreans assumed that the ratio of any two segments is expressible as a ratio of two numbers, i.e., that it is possible to construct a similarity theory based on rational numbers. Thus they tried to reduce geometry to arithmetic.

Also, Pythagoras discovered that the ratios of consonant musical intervals reduce to simple numerical ratios: 1:2 (an octave), 2:3 (a fourth), 3:5 (a fifth), i.e., that the qualitative differences of the sounds of strings can be reduced to the ratios of their lengths, and thus to numbers.

The Pythagoreans concluded that "all is number", that the mathematical relations of magnitudes and the laws of nature are expressible in terms of whole numbers and their ratios. This was the first attempt at arithmetization of mathematics and of mathematical natural science. But this attempt soon collapsed.

A discovery made in the Pythagorean school unsettled all their views. What was proved was that the ratio of the diagonal to the side of a square is not expressible as the ratio of whole numbers. This meant that the whole numbers and their ratios do not suffice for the construction of a similarity theory and of metric geometry.

The only discoveries of comparable significance were the discovery of non-Euclidean geometry in the 19th century and two discoveries made in the first third of the 20th century, namely, the theory of relativity and Gödel's incompleteness theorem. Like these discoveries, the discovery of incommensurability soon became known to most educated people. Plato learned about incommensurability rather late and wrote that before becoming aware of it he resembled an unreasoning animal. Aristotle frequently turned to problems involving incommensurabiliy. He wrote in his *Metaphysics*: "People are at first amazed and ask if things are really that way. Their amazement is comparable to that elicited by toys that move by themselves, by the rotation of the sun, or by the incommensurability of the diagonal, for those who do not know the cause marvel that there is something that cannot be measured by however small a measure" (*Meth.* 983^a).

The discovery of incommensurability proved the unsoundness of the first attempt at arithmetization of mathematics and became the starting point of deep and subtle theories (equivalent to our theory of real numbers, of the classification of irrationalities, and so on). Its immediate consequence was the reversal of the roles of arithmetic and geometry. First the Pythagoreans tried to reduce geometry to arithmetic. But having found out that the realm of geometric magnitudes is larger than that of the rationals they adopted geometry as the basis of mathematics.

2. Geometric algebra

Around the end of the 5th century BCE algebra also put on geometric "armor". The Greeks translated all arithmetical operations into geometric language and began to operate with geometric objects such as segments, areas, and volumes without using numbers. Following H. G. Zeuthen, one usually refers to this period as the period of geometric algebra.

The sources on which we base our evaluation of geometric algebra come from the Hellenistic period (from the 3rd to the 1st centuries BCE) which followed the conquests of Alexander the Great. At that time Hellenistic science resumed contacts with the mathematical and astronomical knowledge of the East. The center of science moved from Asia to Egyptian Alexandria. At the time Egypt was ruled by the Ptolemys, the first of whom—Ptolemy Soter—established in Alexandria the Museon (home of the Muses), a counterpart of an Academy of Science connected with a university, and next to it a splendid library which at one time housed more than 700,000 manuscripts. The greatest Hellenistic scholars were invited to this center. One of the first to come was Euclid.

Between the 3d and 2nd centuries BCE the Alexandrian scientific school was peerless. Here resided geniuses such as Archimedes, Apollonius, Eratosthenes, and the astronomer Hipparchus. True, after a period of training

Archimedes returned to his native Syracuse (in Sicily), but to the end of his life he corresponded with his friends, the Alexandrian mathematicians. Almost all of his works that have come down to us are in the form of letters, each of which is a finished and beautifully written scientific memoir.

In the sequel we will return on many occasions to the algebraic achievements of the Hellenistic mathematicians.

The foundations of geometric algebra are presented in Book II of Euclid's *Elements*. Its propositions are also found in Books VI and XIII. Segments made up the first class of objects of algebra. One could add them (i.e., lay one off next to another) and subtract a smaller from a larger one. The product of two segments was the rectangle on these segments and the product of three segments was the rectangular parallelepiped on them. There was no definition of a product of more than three segments. It made no more sense to speak of a product of more than three segments than it did to speak of spaces of four or five dimensions.

Using geometric algebra it was possible to prove properties of operations and identities known already to the Babylonians. For example, Euclid's *Elements* contains a proof of the distributive law of multiplication over addition. Specifically, Proposition II_1 (i.e., Proposition I in Book II) states that: *If there be two straight lines, and one of them be cut into any number of segments whatever, the rectangle contained by the two straight lines is equal to the rectangles contained by the uncut straight line and each of the segments* (note that by "straight line" Euclid always means a bounded straight line, i.e., a segment). In other words, consider a rectangle on segments a and b: if, say, segment a is divided into parts $a = a_1 + \cdots + a_n$, then $(a_1 + \cdots + a_n)b = a_1 b + \cdots + a_n b$ (Figure 3).

The transition to geometric language, more precisely to constructions realizable by ruler and compass, made it possible for the first time to give general proofs of certain algebraic identities. Thus in II_4 one considered a suitable figure (Figure 4) and proved the identity

$$(a + b)^2 = a^2 + b^2 + 2ab. \tag{1}$$

Here the lengths of the segments a and b are arbitrary, and they can be commensurable or not.

The identity

$$ab = \left(\frac{a+b}{2}\right)^2 - \left(\frac{a-b}{2}\right)^2, \tag{2}$$

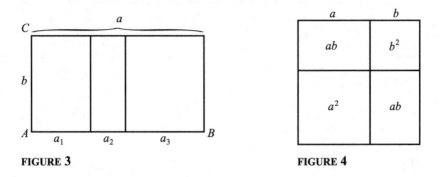

FIGURE 3   **FIGURE 4**

widely used by the ancients in connection with the solution of quadratic equations, was proved with the help of a "gnomon" (Figure 5): if we put

$$AB = a, \qquad BD = b, \qquad AC = CD = \frac{a+b}{2},$$

$$BC = \frac{a-b}{2}, \qquad DF = BD,$$

then the area $ABB'A'$ is equal to the gnomon $CDD'HB'C' = CD^2 - CB^2$, i.e., we obtain the identity (2). Note that the same geometric equality can be expressed differently: putting $\frac{a+b}{2} = u$ and $\frac{a-b}{2} = v$ we obtain

$$(u+v)(u-v) = u^2 - v^2.$$

Problems equivalent to quadratic equations were also given a geometric formulation. Three types of such problems were considered in ancient mathematics.

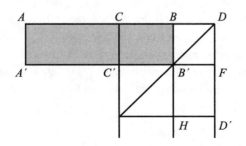

FIGURE 5

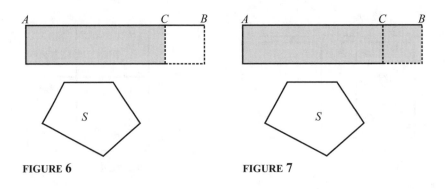

FIGURE 6   FIGURE 7

   1. To transform a given rectangle into a square. This is equivalent to solving the equation $x^2 = ab$.

   2. To "apply" to a given segment $AB = a$ a rectangle of given area S so that the "deficiency" is a square (Figure 6).[1] Putting $BC = x$ we see that this is equivalent to solving the equation

$$x(a - x) = S. \qquad (3)$$

   3. To "apply" to a given segment $AB = a$ a rectangle of given area S so that the "excess" is a square (Figure 7).[1] It is easy to see that this problem is equivalent to solving the equation

$$x(a + x) = S. \qquad (4)$$

   In antiquity the second problem was called elliptic, from the word ἐλλειψις—falling short, and the third hyperbolic, from the word ὑπερβολή—excess. The ancients realized that the second problem was solvable (i.e., had a positive real root) for $S \leq a^2/4$. It is clear that this condition was obtained by finding the maximum of $x(a - x)$ for $0 \leq x \leq a$.

   In Book II of the *Elements* Euclid solves the first of these problems and prepares the technical tools for the solution of the other two, which he solves in generalized form in Book VI. Specifically, instead of rectangles he applies to the given segments parallelograms of given area and requires the deficiency in one case and the excess in the other case to be similar to a given parallelogram. In modern terms, Euclid considers the equations

$$\gamma x(a - x) = S; \qquad (3')$$

$$\gamma x(a + x) = S, \qquad (4')$$

where γ is a positive real number.

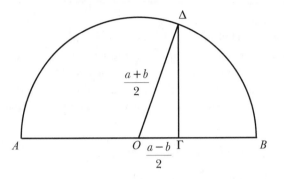

FIGURE 8

Problems 1–3 are solved by first transforming ab, $(a-x)x$, and $(a+x)x$ into a difference of squares (in accordance with formula (2)) and then solving for the unknown by Pythagoras' theorem. Thus in the case of problem 1 one takes a segment AB of length $a+b$, bisects it at O, draws a circle of radius AO, and erects a perpendicular to AB at the end Γ of the segment $A\Gamma = a$. If Δ is its point of intersection with the circle, then the segment $\Gamma\Delta$ yields the side x of the required square (Figure 8).

What was the class of problems solvable by the methods of geometric algebra?

We saw that ruler and compass constructions made possible the solution of various problems equivalent to quadratic equations with real positive roots. It is not difficult to see that they enable one to solve problems reducible to a sequence of quadratic equations in which the coefficients of each equation depend on the roots of its predecessor. The *Elements* contain problems such as the determination of the sides of a regular polygon and of a regular polyhedron in terms of the diameter of its circumscribed circle and sphere respectively. Some of these problems are algebraically equivalent to the solution of a biquadratic equation that is reducible to a succession of two quadratic equations. Thus the expression for the side of a regular pentagon in terms of the diameter of its circumscribed circle is

$$a_5 = R\sqrt{\frac{5}{2} - \frac{\sqrt{5}}{2}},$$

and is therefore reducible to the solution of the pair of quadratic equations

$$y^2 - 5R^2y + 5R^4 = 0; \qquad x^2 = y.$$

In other words, at this stage, the problem of solution of determinate equations by radicals was limited to the solution of such equations in terms of square roots and their combinations. However, problems soon turned up that could not be solved by methods of classical geometric algebra. Before considering such problems we look at how the ancients classified the irrationalities arising in connection with the solution of a chain of quadratic equations.

3. Classification of quadratic irrationalities

The incommensurability of the side and the diagonal of a square was proved in the school of Pythagoras not later than the middle of the 5th century BCE. In algebraic terms, what was proved was that the equation $x^2 = 2$ has neither integral nor rational solutions, or, in modern terms, that $\sqrt{2}$ is an irrational number. One of the oldest proofs, mentioned by Aristotle in his *Prior Analytics* (41a21–41b2), is found in certain copies of Euclid's *Elements*. The proof is by contradiction. Aristotle asserts that if the diagonal and side of a square were commensurable, then "the odd would be equal to the even". O. Becker showed that the proof can be carried out using only the theory of even and odd numbers, which formed the core of Pythagorean arithmetic. (Here is a sketch of the proof. Suppose $\sqrt{2} = m/n$, where m and n are not both even. Then $m^2 = 2n^2 \implies m^2$ even $\implies m$ even $\implies n$ odd, $m = 2t \implies n^2 = 2t^2 \implies n^2$ even $\implies n$ even. Thus n is both even and odd. Contradiction.) In Plato's *Theaetetus* it is told that the Pythagorean Theodorus of Cyrene (second half of the 5th century BCE) presented a lecture to Athenian youths in which he proved that the sides of squares with areas $3, 5, 6, \ldots, 15$ are incommensurable with the side of a unit square. (It is not clear from the dialogue whether Theodorus included the square with area 17 or just went "up to" it. J. Itard showed that, in the latter case, he could have carried out the proof using only the theory of the odd and the even, without having recourse to general divisibility theorems.) In the same dialogue, the young Athenian mathematician Theaetetus stated a general theorem to the effect that the side of a square of area N, N a nonsquare integer, is incommensurable with the side of a unit square, i.e., that if $N \neq \alpha^2$, then $\sqrt{N}$ is not expressible as a rational number. We note that the proof of this theorem requires the use of the general theory of divisibility.

A scholium on Euclid shows that in addition to irrationalities of the form $\sqrt{N}$ Theaetetus considered three other classes of quadratic irrationalities, namely

1) binomials: $M + \sqrt{N}; \quad \sqrt{M} + \sqrt{N};$

2) apotomes (differences): $M - \sqrt{N}$; $\sqrt{M} - \sqrt{N}$;

3) medials (means): $\sqrt{\sqrt{M}\sqrt{N}}$

Euclid extended Theaetetus' classification (his theory of irrationalities and their classification is found in Book X of the *Elements*). It seems that he wanted to construct a domain Ω of numbers closed under the four elementary operations of arithmetic (with limited subtraction) and the operation of extraction of square roots of positive numbers, i.e., a domain in which it is possible to solve every equation of the form

$$x^2 = a^2 + b^2, \qquad a, b \in \Omega.$$

Nowadays we call such a domain a Pythagorean field. All constructions in the *Elements* are carried out over this field (except that subtraction is limited to the cases in which the subtrahend is less than the minuend).

In addition to the classes of irrationalities just listed, Euclid considered the classes $\sqrt{M + \sqrt{N}}, \sqrt{\sqrt{M} + \sqrt{N}}, \sqrt{M - \sqrt{N}}, \sqrt{\sqrt{M} - \sqrt{N}},$ $\sqrt[2^k]{M \cdot N}$. He proved that these classes are nonempty and disjoint.

We note that Euclid's classification has a considerable "reserve", in the sense that there are more classes than needed for the solution of problems considered in the *Elements*. It follows that Euclid attached independent significance to his classification.

According to Plato, Theaetetus defined cubic irrationalities $\sqrt[3]{N}$ as well. It is conceivable that their classification was contained in one of Apollonius' works. Be that as it may, no such classification has come down to us.

4. The first unsolvable problems

In the 5th century BCE three problems were posed that immediately became very famous. These are the problems of duplication of a cube, of trisecting an angle, and of squaring a circle. All of them have a long history. The first two were solved only in the 1830s and the third only at the end of the 19th century. All three turned out to be unsolvable by classical geometric algebra. The nature of the first two of these problems is very different from that of the third.

We will consider in detail only the first of these problems because of its special significance for the evolution of algebra. We can state it as follows: Construct a cube whose volume is twice the volume of a given cube. If a is the side of the given cube and x that of the required cube, then the problem is equivalent to solving the equation $x^3 = 2a^3$.

This problem was so popular that a legend was made up about it. It seems that Athens was afflicted by the plague, and the pronouncement of the oracle at Delos was that the plague would cease if the cubical altar to Apollo were doubled in size. Hence the name Delian problem.

It was natural to try to solve the problem by ruler and compass constructions. Translated into the language of algebra, this meant that one tried to represent $\sqrt[3]{2}$ as a finite combination of quadratic radicals. When these attempts ended in failure the problem was thoroughly investigated. Hippocrates of Chios, who lived and worked in Athens and was one of the best mathematicians of the second half of the 5th century BCE, generalized the problem: Given a rectangular parallelepiped with base a^2 and height b find a cube of the same volume, i.e., solve the equation

$$x^3 = a^2 b. \tag{5}$$

Hippocrates showed that the solution of this equation is equivalent to finding two mean proportionals between a and b, i.e., to solving the equations

$$\frac{a}{x} = \frac{x}{y} = \frac{y}{b}. \tag{6}$$

If $b = 2a$, then x is the side of the required cube.

Archytas of Tarentum (5th century BCE) soon showed that the segment x can be found by considering the intersection of three surfaces of rotation: a cone, a cylinder, and the surface (a "degenerate torus") obtained by rotating a circle about one of its tangents. The ancients had no doubts about the existence of surfaces obtained by rotating a right triangle about one of its legs, of a rectangle about one of its sides, and of a circle about one of its tangents. While Archytas' solution proved the existence of a solution of Hippocrates' problem, it provided no convenient method for the construction of two mean proportionals between two arbitrarily given segments. With a view to finding such a method scholars focussed their attention on equations (6) and on geometric loci, i.e., on curves obtained from Hippocrates' proportion:

$$ay = x^2; \quad xy = ab \quad (\text{or } bx = y^2).$$

The construction of the point of intersection of these curves yielded the solution of the problem. But the investigation of loci was far from simple. First one had to show that these loci were continuous curves, for it was only then that one could talk about the point of their intersection. In the second half of the 4th century BCE Menaechmus, a pupil of Eudoxus, managed to represent these loci as plane sections of cones of rotation. He considered three

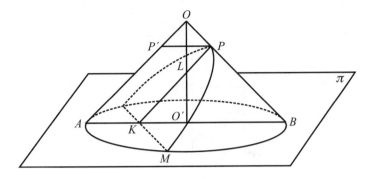

FIGURE 9

cones of rotation, namely right-angled (i.e., having a right angle at the vertex), obtuse-angled, and acute-angled. Sections by planes perpendicular to a given ray lying on each of these three cones yielded three kinds of curves, now known as a parabola, an hyperbola, and an ellipse respectively. These names were introduced by Apollonius (3rd century BCE). Before him these curves were known as sections of right-angled, obtuse-angled, and acute-angled cones. Having given these three-dimensional definitions of the three curves, Menaechmus deduced their plane geometric properties (symptoms) and subsequently operated with these alone. As an example, we show how he obtained the "symptom" of a parabola. Let AOB be the section of a right-angled cone of rotation by a vertical plane passing through its axis OL and let PMK be the trace of a plane perpendicular to its ray OB (Figure 9). Since AMB is a semicircle, $KM^2 = AK \cdot KB$. But $AK = PP' = \sqrt{2LP^2}$ and $KB = \sqrt{2KP^2}$. Hence

$$KM^2 = 2LP \cdot KP.$$

Denote KM by y, KP by x, and LP by p. Then

$$y^2 = 2px. \tag{7}$$

This is the equation (or "symptom") of the curve in modern notation. The ancients expressed it in terms of geometric algebra: at each point of the curve the square on the semichord KM is equal to the rectangle $PKSR$ constructed on the segment PK (from K to the vertex P of the curve) and on the constant segment (parameter) PR (Figure 10).

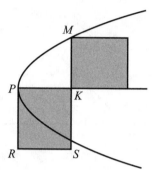

FIGURE 10

The equations

$$\frac{y^2}{x(2a-x)} = \frac{p}{2a} \quad \text{and} \quad \frac{y^2}{x(2a+x)} = \frac{p}{2a},$$

of sections of an acute-angled and obtuse-angled cone respectively, were obtained in a similar manner and were also expressed in terms of geometric algebra. Since the first of these equations is of elliptic type and the second of hyperbolic type, Apollonius called the corresponding curves ellipse and hyperbola respectively. Similarly, the curve defined by equation (7) was called a parabola because this equation is of parabolic type. In other words, Apollonius' classification was algebraic and was based on the application of geometric algebra to geometry. This is very much like our use of literal algebra in modern analytic geometry for the deduction of the equation of a curve and the study of its properties.

We see that investigation of the problem of duplication of a cube resulted in the introduction into mathematics of conic sections, curves of fundamental importance. In antiquity their use was limited to the solution of cubic equations, but two millennia later, in the 17th century, they turned up in Kepler's astronomical laws and played a vital role in mechanics.

We now turn to the second remarkable problem of antiquity. The ancients did not reduce the angle trisection problem to a cubic equation. They solved it either by a so-called *neusis* (verging or inclination) construction or by means of new curves. The first of these methods consists in fitting a line segment of given length between two curves so that it—or its extension—passes through a given point. For example, to trisect a given angle $\angle AOB$ one draws a circle of arbitrary radius R centered at the vertex O, extends OB to the left of O,

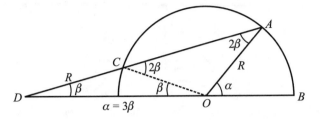

FIGURE 11

and fits a segment CD of length R between the straight line OB and the circle so that its extension passes through A. It is not difficult to see that $\angle CDO = 1/3 \angle AOB$ (Figure 11). But the solution of a problem by a *neusis* construction invites questions. Indeed, if one moves a segment of length a so that one of its endpoints stays on a given curve L and its extension passes through a given point A, then its second endpoint describes a certain curve Γ. A *neusis* construction is equivalent to finding the intersection of the curve Γ with the second given curve. But Γ can be a very complicated curve. Thus if L is a straight line then Γ is a conchoid (its equation in polar coordinates is $\rho = a + b/\cos\varphi$). This is an algebraic curve of 4th degree. (Incidentally, this curve was introduced in the 3rd century BCE.) Some *neusis* constructions can be realized by means of conic sections while others involve curves of higher degree. That is why it was not enough to solve a problem by means of a *neusis* construction. The additional requirement was to explain what curves the construction called for.

The problem of squaring a circle, i.e., of constructing a square whose area is that of a given circle, is not an algebraic problem, but the nature of the first approach to its solution via squarable lunes was algebraic.

A lune is a figure bounded by two arcs of two circles. Hippocrates of Chios found three squarable lunes. We will discuss one of them.

Consider the semicircle AA_1B_1B (Figure 12) with the inscribed isosceles triangle $A\Gamma B$ and the circular segment S_3 similar to the segments $S_1 = AA_1\Gamma$ and $S_2 = \Gamma B_1 B$. Hippocrates showed that the lune $A\Gamma B$ has the same area as the triangle $A\Gamma B$.

His first step was to show that the ratio of similar circular segments, i.e., segments whose central angles are both equal to $(m/n) \cdot 2\pi$, is equal to the ratio of the squares of their chords. In the case under consideration the three central angles are equal to $\pi/2$. Since $AB^2 = 2A\Gamma^2$, it follows that

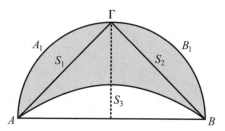

FIGURE 12

$S_3 : S_1 = 2 : 1$. Hence $S_3 = S_1 + S_2$, i.e., the lune $A\Gamma B\Delta = \frown - S_3 = \frown - (S_1 + S_2) = \Delta A\Gamma B$.

Hippocrates wanted to use his lunes to effect the quadrature of a circle. The outer arc of the lune just considered is equal to a semicircle. One of the outer arcs of Hippocrates' two other lunes was greater than a semicircle and the other, smaller than a semicircle. Hippocrates also obtained a lune which, while itself not squarable, was squarable together with a circle. This approximation to a solution of the basic problem was apparent rather than real, a rather frequent occurrence in mathematics. However, the discovery of "Hippocratic lunes" was of independent interest.

In the 18th century the problem of squarable lunes was investigated by D. Bernoulli, G. Cramer, and Euler. They found two more squarable lunes. At the beginning of the 19th century Hippocratic lunes were studied by T. Clausen, a professor at Dorpat (now Tartu) university. He rediscovered the five squarable lunes and advanced the idea that there were no others. The problem was solved in the 20th century by the Bulgarian L. Chakolov and the Russian N. G. Chebotarev and his students. They used Galois theory to show that there are exactly five squarable lunes and none of them is squarable together with a circle.

Another problem that attracted the attention of ancient mathematicians, and was settled only at the end of the 18th century, was the problem of construction of regular polygons. Already the early Pythagoreans knew how to use ruler and compass to inscribe in a circle regular n-gons for $n = 3, 4, 5$ as well as all regular polygons obtained from the latter by doubling the number of their sides. In addition, Euclid's *Elements* contains a construction of a regular polygon with 15 sides. In particular, the ancients were interested in the case $n = 7$. What has come down to us is an Arabic translation of a work of Archimedes in which he showed how to divide a circle into seven equal parts. Archimedes used a *neusis* construction that could be realized by means of conic sections.

The problem of construction of regular polygons by means of ruler and compass was settled at the end of the 18th century by the young Gauss, who reduced it to the solution of the cyclotomic equation $x^n - 1 = 0$. We will discuss this matter in the sequel. The unsolvability of the first two famous problems of antiquity was proved in 1837 by P. L. Wantzel. But already in Euclid's time mathematicians were inclined to think that all three famous problems were unsolvable by means of ruler and compass. At any rate, these problems were not included in Euclid's *Elements*, in which the basic construction tools are ruler and compass.

The first classification of problems dates back to Euclid's time (or may have been carried out somewhat later). Problems belonged to one of three classes. The first class was the class of "plane" problems, solvable by ruler and compass. The second was the class of "solid" problems, solvable by conic sections. All remaining problems formed the class of "linear" problems. Pappus (3rd century CE) considered it a major error when a "plane" problem was solved by means of conic sections. In Archimedes' hands the latter became the universal means of solution of cubic equations.

5. Cubic equations

A more systematic investigation of problems equivalent to cubic equations took place in the Hellenistic period (3rd–2nd centuries BCE). In *On the Sphere and Cylinder* (see *The Works of Archimedes*, ed. T. L. Heath, Dover.), Archimedes reduced the problem of dividing a sphere by a plane into two segments with volumes V_1 and V_2 in a prescribed ratio, $V_1 : V_2 = m : n \ (m > n)$, to finding the height x of the larger segment from the proportion

$$4R^2 : x^2 = (3R - x) : \frac{m}{m+n},\tag{8}$$

where R is the radius of the sphere.

Archimedes replaced this problem by the following generalized version: To divide a given segment a into two parts x and $a - x$ such that

$$(a - x) : c = S : x^2,\tag{9}$$

where c and S are a given segment and a given area respectively.

Archimedes realized that the generalized version of the initial problem was not always solvable (i.e., it did not always admit a positive solution). This, he said, meant that one had to impose restrictions ("diorisms") on the data, and promised to carry out an analysis and a synthesis. But the promised solution got lost. It was not mentioned in any of the lists of the works of Archimedes. The Greek geometers Diocles and Dionisidorus, who lived a century after Archimedes and provided rather involved solutions of the original problem rather than of its generalized version, seemed to be unaware of it. It was not until the 6th century CE that the Archimedes commentator Eutocius found the lost solution, now usually added as an appendix to the collected works of Archimedes. What follows is the solution in question.

Archimedes puts $S = pb$ and solves the problem by considering two conic sections: the parabola

$$y = x^2/p \tag{10}$$

and the hyperbola

$$y = cb/(a - x). \tag{11}$$

Both of these equations are easily obtained from the proportion (9). It is remarkable that Archimedes goes over from the proportion (9) to the cubic equation

$$x^2(a - x) = cS, \tag{12}$$

which he expresses in words as a relation between volumes. He needs this transition in order to be able to clarify the conditions for the existence of a solution. It is clear that the condition that assures the existence of positive roots of equation (12) is

$$cS \leq \max x^2(a - x), \qquad 0 \leq x \leq a.$$

In other words, the problem reduces to finding the maximum of $x^2(a - x)$.

Archimedes claims that the maximum in question is attained for $x = \frac{2}{3}a$ and proves his claim. Without discussing his infinitesimal arguments we state his conclusion that if $cS < \max x^2(a-x) = \frac{4}{27}a^3$, then equation (12) has two different positive real roots (Figure 13, case I) which can be obtained from the points of intersection of the curves (10) and (11). If $cS = \frac{4}{27}a^3$, then these curves have a common tangent (Figure 13, case II), i.e., the equation has a double root. Finally, if $cS > \frac{4}{27}a^3$, then the curves (10) and (11) do not intersect for $x \in [0, a]$ (Figure 13, case III).

It was not until the 19th century that one encountered an equally complete and profound analysis of a problem.

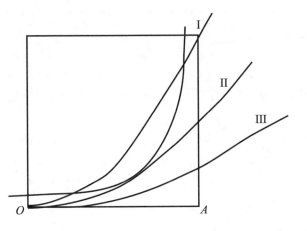

At the end of his letter to Dosipheus, which is a kind of introduction to his *On Conoids and Spheroids*, Archimedes notes that the theorems proved in this work enable one to solve many problems, such as, for example, the following: "From a given spheroid (ellipsoid of rotation) or conoid (paraboloid or hyperboloid of rotation) to cut off, by means of a plane parallel to a given plane, a segment equal to a given cone, cylinder, or sphere." These problems, as well as the problems of division of a sphere, reduce to cubic equations. In the case of an obtuse-angled conoid the equation is of the form

$$x^2(a \pm x) = cS.$$

Archimedes' letter makes it clear that he analyzed and solved this equation. Thus he may be said to have considered cubic equations of the form $x^3 + ax^2 = b$ for different values of a and b and to have given methods for their solution. In antiquity the investigation and solution of cubic equations in general form presented difficulties that only Archimedes managed to overcome. Using the new geometric machinery of conic sections, Greek mathematicians solved particular problems equivalent to cubic equations. This method was subsequently adopted by mathematicians in Islamic countries who tried to analyze all cubic equations.

In summary, we can say that the roots of cubic equations, including the roots of the equation of duplication of a cube, were determined by intersecting parabolas and hyperbolas. In antiquity no one posed the problem of solution of equations in terms of cubic radicals, leave alone in terms of radicals of higher degree. At this stage the geometric solution of cubic equations had little effect on the evolution of algebra. Had geometric algebra continued to develop it would have, most likely, led to the investigation of curves of higher degree. However, as we will see in the sequel, the development of geometric algebra was interrupted in the early centuries CE and algebra followed a different path.

6. Indeterminate equations

In the theoretical works from the 5th–3rd centuries BCE we find just two types of indeterminate equations, namely the Pythagorean equation

$$x^2 + y^2 = z^2 \tag{13}$$

and the equation

$$y^2 - ax^2 = \pm 1, \quad a \neq \alpha^2, \tag{14}$$

which later came to be known as the Pell-Fermat equation. In both cases one looked for positive integral solutions.

The theme of the Pythagorean equation runs through all of ancient mathematics. The early Pythagoreans found a general solution in the form

$$x = a, \qquad y = \frac{a^2 - 1}{2}, \qquad z = \frac{a^2 + 1}{2},$$

where a is an odd number. The simplest solution is the triple $(3, 4, 5)$. Another general solution was given by Plato, who started with an even a and put

$$x = a, \quad y = \left(\frac{a}{2}\right)^2 - 1, \qquad z = \left(\frac{a}{2}\right)^2 + 1.$$

For $a = 4$ we again get the triple $(3, 4, 5)$.

Neither of these two general solutions of equation (13) is complete. This is clear from the fact that each of them involves a single parameter whereas the complete general solution must involve two parameters.

A complete general solution of (13) is found in Book X of Euclid's *Elements* (Proposition 29, Lemma 1). In modern terms, this solution can be written as follows:

$$x = p^2 - q^2, \qquad y = 2pq, \qquad z = p^2 + q^2.$$

The early Pythagoreans also considered the special case of equation (14) with $a = 2$. Let us rewrite this equation as

$$\frac{y^2}{x^2} - 2 = \pm \frac{1}{x^2}. \tag{15}$$

If we put for y and x values x_n and y_n obtained from the recursive formulas

$$x_n = x_{n-1} + y_{n-1};$$
$$y_n = 2x_{n-1} + y_{n-1},$$

(16)

$n = 1, 2, \ldots$ and $x_0 = y_0 = 1$, then we see that with increasing x_n the ratios $y_n : x_n$ come ever closer to $\sqrt{2}$; they are ever better "rational diagonals" of the unit square. In the dialogue *The State*, Plato calls the number seven a "rational diagonal" of the square with side five. By way of an explanation Proclus writes that one took ones as the first values of the diagonal and the side ($x_0 = 1$, $y_0 = 1$ give the least solution of equation (14) for $a = 2$), and that the subsequent values were obtained from the formulas (16) for successive values of n. Then $x_1 = 2$, $y_1 = 3$, $x_2 = 5$, $y_2 = 7$, i.e., the second step yields Plato's "rational diagonal".[2]

According to Proclus, Propositions II_9 and II_{10} of Euclid's *Elements* prove that formulas (16) give the successive solutions of equation (14) for $a = 2$. Indeed, Proposition II_9 (in which a segment AB is cut into equal segments at C and into unequal segments at D; see Figure 14) gives a geometric proof of the identity

$$AD^2 + BD^2 = 2AC^2 + 2CD^2,$$

where $AC = CB$, or, equivalently, $AD^2 - 2AC^2 = -(BD^2 - 2CD^2)$. Thus if $BD^2 - 2CD^2 = \pm 1$, then $AD^2 - 2AC^2$ has that same value. But, as is clear from Figure 14 (we are putting $CD = x$ and $BD = y$),

$$AC = CD + BD = x + y;$$
$$AD = 2CD + BD = 2x + y.$$

Proposition II_{10} is dual to Proposition II_9. It concerns the case when D lies on the extension of the segment AB.

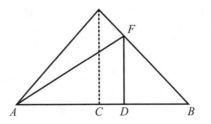

FIGURE 14

It is very likely that in Archimedes' time mathematicians knew formulas for finding infinitely many solutions of equation (14), starting with the smallest one, for values of a other than 2. An argument in favor of this assumption is that in *Measurement of a Circle* Archimedes gives an approximation to $\sqrt{3}$ that can be obtained from the recursion formulas

$$x_n = x_{n-1} + y_{n-1}; \qquad y_n = 3x_{n-1} + y_{n-1}.$$

Specifically, starting with $x_0 = y_0 = 1$, we get the following sequence of approximations to $\sqrt{3}$:

$$\frac{1}{1}, \quad \frac{2}{1}, \quad \frac{5}{3}, \quad \frac{7}{4}, \quad \frac{19}{11}, \quad \frac{26}{15}, \quad \frac{71}{41}, \quad \frac{97}{56}, \quad \frac{265}{153}, \quad \cdots$$

Archimedes used the approximation $\frac{265}{153}$ without explaining its origin. This indicates that the method he used to obtain it was widely known.

We also know that Archimedes set for the Alexandrian mathematicians the so-called "problem about bulls", which reduces to the solution of equation (14) for $a = 4729494$. It seems that what the great mathematician wanted to know was not a concrete solution of equation (14) for a given a but whether the Alexandrian mathematicians had a general method for getting the relevant least solution as well as the recursion relations for obtaining the remaining solutions.

The subsequent history of indeterminate equations indicates that during the period under consideration many other such equations were solved, but their investigation was not part of the theoretical science of that time.

Between the time when this essay was written and the present moment the authors' views concerning the origin of geometric algebra have changed: geometric algebra seems to have been a *universal* language of mathematics in antiquity and during the Middle Ages. In this connection see their paper "A new view of the geometric algebra of the ancients" in the Appendix at the end of the book.

Editor's notes

[1] A less condensed version of this statement would begin as follows: To construct on a given segment a rectangle whose area equals that of a given area S so that. . .

2 The discussion in 2 of Chapter I describes how a composition formula was used in Babylonian mathematics. Here, a similar composition formula can be used to explain the recursive formulas (16).

Equation (15) can be written as $y^2 - 2x^2 = \pm 1$, and solutions of this equation can be found using the composition formula

$$(y^2 - 2x^2)(s^2 - 2r^2) = (2rx + sy)^2 - 2(sx + ry)^2 \qquad (*)$$

as follows. Put $y = y_{n-1}$, $x = x_{n-1}$, $r = 1, s = 1$ in $(*)$. Using (16), this gives $-(y_{n-1}^2 - 2x_{n-1}^2) = y_n^2 - 2x_n^2$. Hence $y_{n-1}^2 - 2x_{n-1}^2 = \pm 1$ implies $y_n^2 - 2x_n^2 = -(\pm 1)$.

CHAPTER **3**

The birth of literal algebra

1. Mathematics in the first centuries CE. Diophantus

The third—very important—stage of the development of algebra began in the first centuries CE and came to an end at the turn of the 17th century. Its beginning was marked by the introduction of a *literal symbolism* by Diophantus of Alexandria, and its end by the creation of a *literal calculus* in the works of Viète and Descartes. It was then that algebra acquired its own distinctive language which we use today.

The first century BCE was a period of Roman conquests and of Roman civil wars. Both were taking place in the territories of the Hellenistic states and the Roman provinces and were accompanied by physical and economic devastation. One after another these states lost their independence. The last to fall was Egypt (30 BCE). The horrors of war and the loss of faith in a secure tomorrow promoted the spread of religious and mystical teachings and undermined interest in the exact sciences and in abstract problems in mathematics and astronomy. In Cicero's dialogue *On the State* one of the participants proposes a discussion of why two Suns were seen in the sky. But the topic is rejected, for "even if we acquired profound insight into this matter, we would not become better or happier."

In the second half of the first century BCE mathematical investigations came to a virtual halt and there was an interruption in the transmission of the scientific tradition.

At the beginning of the new era economic conditions in the Hellenistic countries, now turned Roman provinces, gradually improved and there was a revival of literature, art, and science. In fact, the second century came to be known as the Greek Renaissance. It was the age of writers such as Plutarch and Lucian and of scholars such as Claudius Ptolemy.

Alexandria continued its role as the cultural and scientific center of antiquity, and in this respect Rome was never its rival. Nor did Rome ever develop an interest in the depths of Hellenistic science. As noted by Cicero in his *Tusculanae disputationes*, the Romans, unlike the Greeks, did not appreciate geometry; just as in the case of arithmetic, they stopped at a narrow, practical knowledge of this subject. They had little regard for all of mathematics. Even accounting, surveying, and astronomical observations were left to the Greeks, the Syrians, and other conquered nations. According to Vergil, the destiny of the Romans was wise government of the world.

The revival of the Alexandrian school was accompanied by a fundamental change of orientation of its mathematical research. During the Hellenistic period geometry was the foundation of Greek mathematics; algebra had not, as yet, become an independent science but developed within the framework of geometry, and even the arithmetic of whole numbers was constructed geometrically. Now number became the foundation. This resulted in the arithmetization of all mathematics, the elimination of geometric justifications, and the emergence and independent evolution of algebra.

We encounter the return to numerical algebra already in the works of the outstanding mathematician, mechanician, and engineer Heron of Alexandria (1st century CE). In his books *Metrica, Geometrica*, and others, books that resemble in many respects our handbooks for engineers, one finds rules for the computation of areas and volumes, solutions of numerical quadratic equations, and a number of interesting problems that reduce to indeterminate equations. In particular, they contain the famous "Heron formula" for the computation of the area of a triangle given its sides a, b, c:

$$S = \sqrt{p(p-a)(p-b)(p-c)},$$

where $p = (a + b + c)/2$. Here the expression under the square root sign is a product of four segments, and thus an expression totally inadmissible in geometric algebra. It is clear that Heron thought of segments as numbers whose products are likewise numbers.

In his famous book, known under its Arabized name *Almagest*, Claudius Ptolemy, when computing tables of chords, identified ratios of magnitudes with numbers and the operation of "composition" of ratios—defined in Euclid's *Elements*—with ordinary multiplication.

The new tendencies found their clearest expression in the works of Diophantus of Alexandria, who founded two disciplines: algebra and Diophantine analysis.

We know next to nothing about Diophantus himself. On the basis of certain indirect remarks, Paul Tannery, the eminent French historian of mathematics, concluded that Diophantus lived in the middle of the 3rd century CE. On the other hand, Renaissance scholars who discovered Diophantus' works supposed that he lived at the time of Antoninus Pius, i.e., approximately in the middle of the 2nd century. An epigram in *The Greek Anthology* provides the following information: "Here you see the tomb containing the remains of Diophantus, it is remarkable: artfully it tells the measures of his life. God granted him to be a boy for the sixth part of his life, and adding a twelfth part to this, He clothed his cheeks with down; He lit him the light of wedlock after a seventh part, and five years after his marriage He granted him a son. Alas! late-born wretched child; after attaining the measure of half his father's life, chill Fate took him. After consoling his grief by this science of numbers for four years he ended his life. By this device of numbers tell us the extent of his life." A simple computation shows that Diophantus died at the age of 84 years. This is all we know about him.

2. Diophantus' *Arithmetica*. Its domain of numbers and symbolism

Only two (incomplete) works of Diophantus have come down to us. One is his *Arithmetica* (six books out of thirteen; four more books in Arabic, attributed to Diophantus, were found in 1973. They will be discussed in the sequel). The other is a collection of excerpts from his treatise *On Polygonal Numbers*. We will only discuss the first of these works.

Arithmetica is not a theoretical work resembling Euclid's *Elements* or Apollonius' *Conic Sections* but a collection of (189) problems, each of which is provided with one or more solutions and with relevant explanations. At the beginning of the first book there is a short algebraic introduction, which is basically the first account of the foundations of algebra. Here the author constructs the field of rational numbers, introduces literal symbolism, and rules for operating with polynomials and equations.

Already Heron regarded positive rational numbers as legitimate numbers (in classical ancient mathematics "number" denoted a collection of units, i.e., a natural number). While Diophantus defined a number as a collection of units, throughout *Arithmetica* he called every positive rational solution of a problem "number" ($\alpha\rho\iota\theta\mu\acute{o}\varsigma$), i.e., he extended the notion of number to all of $\mathbf{Q}^+$. But this was not good enough for the purposes of algebra, and so Diophantus took the next decisive step of introducing negative numbers. It was only then that he obtained a system closed under the four operations of algebra, i.e., a field.

How did Diophantus introduce these new objects? Today we would say that he used the axiomatic method: he introduced a new object called "deficiency" ($\lambda\varepsilon\tilde{\iota}\psi\iota\varsigma$, from $\lambda\varepsilon\tilde{\iota}\pi\omega$—to lack) and stated rules for operating with it. He writes: "deficiency multiplied by deficiency yields availability (i.e., a positive number); deficiency multiplied by availability yields deficiency; and the symbol for deficiency is ⋔, an inverted and shortened (letter) ψ" (Diophantus. *Arithmetica*. Definition IX). In other words, he formulated the rule of signs which we can write as follows:

$$(-) \times (-) = (+),$$
$$(-) \times (+) = (-).$$

Diophantus did not formulate rules for addition and subtraction of the new numbers but used them extensively in his books. Thus, while solving problem III_8 (i.e., Problem 8 in Book III), he needs to subtract $2x + 7$ from $x^2 + 4x + 1$. The result is $x^2 + 2x - 6$, i.e., here he carries out the operation $1 - 7 = -6$. In problem VI_{14}, $90 - 15x^2$ is subtracted from 54 and the result is $15x^2 - 36$. Thus here $-15x^2$ is subtracted from zero; in other words, Diophantus is using the rule $-(-a) = a$.

We note that Diophantus used negative numbers only in intermediate computations and sought solutions only in the domain of positive rational numbers. A similar situation developed later in connection with the introduction of complex numbers. Initially they were regarded as just convenient symbols for obtaining results involving "genuine", i.e., real, numbers.

Diophantus also introduced literal signs for an unknown and its powers. He called an unknown a "number" ($\dot{\alpha}\rho\iota\theta\mu\acute{o}\varsigma$) and denoted it by the special symbol ς. It is possible that this symbol was introduced before him. We find it in the Michigan papyrus (2nd century CE) as well as in a table appended to Heron's *Geometrica*. But Diophantus boldly breaks with geometric algebra by introducing special symbols *for the first six positive powers of the unknown, the first six negative powers, and for its zeroth power*. While the square and cube of the unknown could be interpreted geometrically, its 4th, 5th, and 6th powers could not be so represented. Nor could the negative powers of the unknown.

Diophantus denoted the positive powers of the unknown as follows:

$$x\text{—}\varsigma; \quad x^2\text{—}\Delta^{\upsilon}; \quad x^3\text{—}K^{\upsilon}; \quad x^4\text{—}\Delta^{\upsilon}\Delta; \quad x^5\text{—}\Delta K^{\upsilon}; \quad x^6\text{—}K^{\upsilon}K.$$

He defined negative powers as inverses of the corresponding positive powers and denoted them by adding to the exponents of the positive powers the symbol χ. For example, he denoted $x^{-2} = 1/x^2$ by $\Delta^{\upsilon\chi}$.

He denoted the zeroth power of the unknown by the symbol $\overset{\circ}{\text{M}}$, that is, by the first two letters in Μόνάς, or unity.

Then he set down a "multiplication table" for powers of the unknown that can be briefly written down as follows:

$$x^m x^n = x^{m+n}, \quad -6 \le m+n \le 6.$$

He singled out two rules that correspond to basic axioms which we use for defining a group:

$$x^m \cdot 1 = x^m \quad \text{(definition VII)}; \tag{1}$$

$$x^m x^{-m} = 1 \quad \text{(definition VI)}. \tag{2}$$

In addition, Diophantus used the symbol $\overset{\prime}{\iota}\sigma$ for equality and the symbol $\square$ for an indeterminate square. All this enabled him to write equations in literal form. Since he did not use a symbol for addition, he first set down all positive terms, then the minus sign (i.e., ⋔), then the negative terms. For example, the equation

$$x^3 - 2x^2 + 10x - 1 = 5$$

was written as

$$\text{K}^{\upsilon}\overline{\alpha\varsigma\iota} \ ⋔ \ \Delta^{\upsilon}\overline{\beta}\overset{\circ}{\text{M}}\overline{\alpha}\overset{\prime\prime}{\iota}\sigma\overset{\circ}{\text{M}}\overline{\epsilon}.$$

Here $\overline{\alpha} = 1$, $\overline{\iota} = 10$, $\overline{\beta} = 2$, $\overline{\epsilon} = 5$ (we recall that the Greeks used the letters of the alphabet to denote numbers).

In the "Introduction" Diophantus formulated two basic rules of transformation of equations: 1) the rule for transfer of a term from one side of an equation to the other with changed sign and 2) reduction of like terms. Later, these two rules became well known under their Arabized names of *al-jabr* and *al-muqābala*.

Diophantus also used the rule of substitution in a masterly way but never formulated it.

We can say that in the introduction Diophantus defined the field **Q** of rational numbers, introduced symbols for an unknown and its powers, as well as symbols for equality and for negative numbers.

Before discussing the contents of *Arithmetica* we consider the possibilities and limitations of Diophantus' symbolism. Getting ahead of the story, we can say that, basically, Diophantus considered in his work indeterminate equations. Such equations always involve two or more unknowns. But he introduced symbols for just one unknown and its powers. How did he proceed when solving problems?

First he stated each problem in general form. For example: "To decompose a square into a sum of squares" (problem II$_8$). Now we would write this problem as

$$x^2 + y^2 = a^2.$$

How could Diophantus write this equation with just one symbol for an unknown and without symbols for parameters (in this case a)? He proceeded as follows: After the general formulation he assigned concrete values to the parameters—in the present case he put $a^2 = 16$. Then he denoted one unknown by his special symbol (we will use the letter t instead) and expressed the remaining unknowns as linear, quadratic, or more complex rational functions of that unknown and of the parameters. In the present example, one unknown is denoted by t and the other by $kt - a$ or, as Diophantus puts it, "a certain number of t's minus as many units as are contained in the side of 16", i.e., instead of a he takes 4 and instead of the parameter k—the number 2. But by saying "a certain number of t's" he indicates that the number 2 plays the role of an arbitrary parameter. Thus Diophantus' version of our equation is

$$t^2 + (2t - 4)^2 = 16,$$

so that

$$x = t = 16/5; \qquad y = 2t - 4 = 12/5.$$

One might think that Diophantus was satisfied with finding a single solution. But this is not so. In the process of solving problem III$_{19}$ he finds it necessary to decompose a square into two squares. In this connection he writes: "We know that a square can be decomposed into a sum of squares in infinitely many ways."

The use of a concrete number to denote an arbitrary parameter has the virtue of simplicity. Sometimes it turned out that the parameter could not be selected arbitrarily, that it had to satisfy additional conditions. In such cases Diophantus determined these conditions. Thus problem VI$_8$ reduces to the system

$$x_1^3 + x_2 = y^3, \qquad x_1 + x_2 = y.$$

Diophantus puts $x_2 = t$, $x_1 = \beta t$, where $\beta = 2$. Then from the second equation we obtain $y = (\beta + 1)t$, and from the first,

$$t^2 = \frac{1}{(\beta + 1)^3 - \beta^3}.$$

Since $\beta = 2$, $t^2 = 1/19$, i.e., t is not rational. In order to obtain a rational solution Diophantus looks at the way t^2 is expressed in terms of the parameter β. The expression in question is a fraction whose numerator, 1, is a square. But then the denominator must also be a square:

$$(\beta + 1)^3 - \beta^3 = \square.$$

Diophantus takes as the new unknown $\beta = \tau$ (he denotes it by the same symbol as the original unknown x_2) and obtains

$$3\tau^2 + 3\tau + 1 = \square.$$

Solving this equation by his method (which we will describe in detail in the next section) Diophantus obtains

$$\tau = \frac{3 + 2\lambda}{\lambda^2 - 3},$$

i.e., the parameter could only be chosen from the class of numbers $\{(3 + 2\lambda)/(\lambda^2 - 3)\}$. Diophantus takes $\lambda = 2$ and obtains $\beta = 7$. Then he goes back to solving the original problem.

Frequently Diophantus deliberately chooses for parameters numbers that do not lead to solutions. He does this in order to show how to analyze problems.

Thus concrete numbers play two roles in *Arithmetica*. One role is that of ordinary numbers and the other is that of symbols for arbitrary parameters. Numbers were destined to play the latter role almost to the end of the 16th century.

Time to sum up. Diophantus was the first to reduce determinate and indeterminate problems to equations. We may say that he did for a large class of problems of arithmetic and algebra what Descartes was later to do for problems of geometry, namely he reduced them to setting up and solving algebraic equations. Indeed, in order to solve problems—arithmetical in the case of Diophantus and geometric in the case of Descartes—both set up algebraic equations which they subsequently transformed and solved in accordance with the rules of algebra. Also, the transformations involved—such as elimination of unknowns, substitutions, and reduction of similar terms—had no direct arithmetic or geometric significance and were not subject to extensive interpretations. In both cases such interpretations were reserved for the final results. We are used to associating this important step with Descartes' creation of analytic geometry, but long before Descartes this step was taken by Diophantus in his *Arithmetica*. None of the scholars of the period between the 13th and 16th centuries unfamiliar with *Arithmetica* entertained the idea of applying algebra to the solution of number-theoretic problems.

3. The contents of *Arithmetica*. Diophantus' methods

We have already mentioned that *Arithmetica* is a collection of problems with solutions. This may create the impression that it is not a theoretical work. But a careful reading makes it clear that the purpose of the painstaking choice and deliberate placement of problems was to illustrate the application of specific general methods. It is a characteristic of ancient mathematics that methods were not formulated apart from problems but were disclosed in the process of their solution. We recall that the famous "method of exhaustion"—the first variant of the theory of limits—was not set down in pure form either by its author Eudoxus of Cnidus or by Archimedes. It was mathematicians of the 16th and 17th centuries who isolated it by analyzing Euclid's *Elements* and Archimedes' quadratures and formulated it in general terms. The same applies to Diophantus' *Arithmetica*. As we show in the sequel, his methods were isolated in the 16th and 17th centuries by Italian and French mathematicians. Following them, we will try to isolate some of these methods and state them in general form.

In Book I Diophantus solves particular determinate linear and quadratic equations. The remaining books deal with the solution of indeterminate equations, i.e., equations of the form

$$F(x_1, \ldots, x_n) = 0, \quad n \geq 2,$$

$F(x_1, \ldots, x_n)$ a polynomial, or of systems of such equations:

$$\begin{cases} F_1(x_1, \ldots, x_n) = 0, \\ \ldots \\ F_k(x_1, \ldots, x_n) = 0, \quad k < n. \end{cases}$$

Diophantus looks for positive rational solutions $x_1^0, x_2^0, \ldots, x_n^0$, $x_i^0 \in \mathbf{Q}^+$, of such equations or of such systems.

It is clear that to solve his determinate equations Diophantus needed only symbols for x and x^2 and not for all x^n, $-6 \leq n \leq 6$. In other words, he extended his domain of numbers and introduced most of his symbols to investigate and solve *indeterminate* equations, where he really needed higher powers of the unknown as well as its negative powers.

Thus the birth of literal algebra was connected not with determinate but with indeterminate equations.

Here we will present just one of Diophantus' methods, namely his method for finding rational solutions of a quadratic equation in two unknowns:

$$F_2(x, y) = 0, \tag{3}$$

where $F_2(x, y)$ is a quadratic polynomial with rational coefficients.

Basically, Diophantus proves the following theorem: If equation (3) has a rational solution (x_0, y_0) then it has infinitely many such solutions (x, y), and x and y are both rational functions (with rational coefficients) of a single parameter:[1]

$$x = \varphi(k), \quad y = \psi(k). \tag{4}$$

When presenting his methods we will use modern algebraic symbolism. This is by now a standard procedure in historico-mathematical literature.

Diophantus began by considering quadratic equations of the form

$$y^2 = ax^2 + bx + c, \qquad a, b, c \in \mathbf{Q}, \tag{5}$$

and put $c = m^2$ (in other words, he assumed that the equation had two rational solutions $(0, m)$ and $(0, -m)$). To find solutions he made the substitution[2]

$$y = kx \pm m \tag{6}$$

and obtained

$$x = \frac{b \mp 2km}{k^2 - a}, \qquad y = k\frac{b \mp 2km}{k^2 - a} \pm m.$$

By assigning to k all possible rational values (Diophantus took only values that yielded positive x and y) we obtain infinitely many solutions of equation (5).

We mentioned earlier problem II_8, which reduces to the equation

$$x^2 + y^2 = a^2, \tag{7}$$

and recall that Diophantus solved it by making the substitution

$$x = t; \qquad y = kt - a, \tag{8}$$

and obtained (we are replacing his numerical values by appropriate letters)

$$x = t = a\frac{2k}{1+k^2}; \qquad y = a\frac{k^2-1}{k^2+1}.$$

To see the sense of this solution and to appreciate its generality we must look at its geometric interpretation. Equation (7) determines a circle of radius a centered at the origin, and the substitution (8) is the equation of a straight line with slope k passing through the point $A(0,-a)$ on that circle (Figure 15). It is clear that the straight line (8) intersects the circle (7) in another point B with rational coordinates. Conversely, if there is a point B_1 with rational coordinates (x_1, y_1) on the circle (7) then AB_1 is a straight line of the pencil (8) with rational slope k. Thus to every rational k there corresponds a rational point on the circle (7), and to every rational point on the circle (7) different from $(0, a)$ there corresponds a rational value of k. Hence Diophantus' method yields all rational solutions of equation (7).

This argument shows that a conic with a rational point is birationally equivalent to a rational straight line.

Next Diophantus considered the more general case when equation (5) has a rational point but the coefficient c is not a square. He first considered

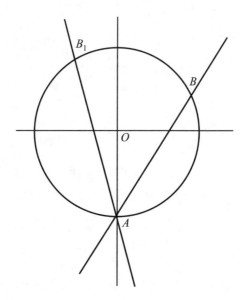

FIGURE 15

this case in problem II_9, which reduces to the equation

$$x^2 + y^2 = a^2 + b^2 \tag{9}$$

(Diophantus put $a = 2, b = 3$). It is clear that equation (9) has the following four solutions: (a, b), $(-a, b)$, $(a, -b)$, and $(-a, -b)$. He makes the substitution

$$x = t + a, \quad y = kt - b \tag{10}$$

and obtains $t = 2[(bk - a)/(1 + k^2)]$. Applying a geometric interpretation analogous to the one just used we see that, essentially, he is constructing a straight line with slope k through $(a, -b)$ on the circle (9).

Diophantus considered a more general case in lemma 2, proposition VI_{12}, and in the lemma for proposition VI_{15}: assuming that equation (5) has a rational solution (x_0, y_0) he made the substitution $x = t + x_0$ and obtained

$$y^2 = at^2 + (2ax_0 + b)t + y_0^2,$$

i.e., he reduced the problem to the case $c = m^2$.

Finally, he considered equation (5) in the case when $a = \alpha^2$. He made the substitution[2]

$$y = \alpha x + k \tag{11}$$

and obtained

$$x = \frac{c - k^2}{2\alpha k - b}.$$

This case calls for a separate discussion. To understand why the straight line (11) intersects the conic section (5) in just one point we introduce projective coordinates (U, V, W) by putting $x = U/W$, $y = V/W$, i.e., we consider our conic in the projective plane $\mathbf{P}^2$. Then equation (5) takes the form

$$V^2 = \alpha^2 U^2 + bUW + cW^2. \tag{12}$$

The curve L so defined intersects the line at infinity $W = 0$ in two rational points: $(1, \alpha, 0)$ and $(1, -\alpha, 0)$. The straight line (11), whose equation in projective coordinates is

$$V = \alpha U + kW,$$

passes through the first of these points.

In summary, we can say that Diophantus carried out a complete investigation of a quadratic indeterminate equation in two unknowns. Later, his analysis served as a model for the investigation of the issue of rational points on curves of genus 0.

Diophantus used more complex and more sophisticated methods to solve equations of the form

$$y^2 = ax^3 + bx^2 + cx + d,$$

$$y^3 = ax^3 + bx^2 + cx + d,$$

$$y^2 = ax^4 + bx^3 + cx^2 + dx + f,$$

and systems of the form

$$\begin{cases} ax^2 + bx + c = y^2, \\ a_1 x^2 + b_1 x + c_1 = z^2, \end{cases}$$

which he called "double equalities". Readers interested in getting a deeper understanding of Diophantus' methods should consult the books [1] and [2] (the latter in Russian) which contain further references to the literature. The history of Diophantus' methods extends all the way to the papers of Poincaré that appeared at the beginning of the 20th century. It was on the basis of these methods that Poincaré constructed an arithmetic of algebraic curves—an area of mathematics that is being intensively developed at the present time.

We conclude our survey by considering Diophantus' problem III_{19}. This problem reduces to a system of 8 equations in 12 unknowns:

$$\begin{cases} (x_1 + x_2 + x_3 + x_4)^2 + x_i = y_i^2, \\ (x_1 + x_2 + x_3 + x_4)^2 - x_i = z_i^2; \quad i = 1, 2, 3, 4. \end{cases}$$

Diophantus notes that "in every right triangle the square of the hypotenuse remains a square if we add to it, or subtract from it, twice the product of its legs." This observation enables him to find a solution using four right triangles with the same hypotenuse. Indeed, let the sides of the four triangles be a_i, b_i, c, $i = 1, 2, 3, 4$. Then it suffices to put $x_1 + x_2 + x_3 + x_4 = ct$, $x_i = 2a_i b_i t^2$, $i = 1, \ldots, 4$, where t is determined by the equation $2a_1 b_1 t^2 + \cdots + 2a_4 b_4 t^2 = ct$. Thus the problem reduces to finding a number c that can be written as a sum of two squares in four different ways. Diophantus solves this essentially number-theoretic problem as follows: he takes two right triangles with respective sides $3, 4, 5$ and $5, 12, 13$ and multiplies the sides of each of them by the hypotenuse of the other. As a result he obtains two right triangles with the same hypotenuse: $39, 42, 65$ and $25, 60, 65$. Now $5 = 1^2 + 2^2$ and $13 = 2^2 + 3^2$. Using the rule for composition of forms $u^2 + v^2$, known already to the Babylonians (see Ch. 1, (15)), he obtains

$$65 = 5 \cdot 13 = (1^2 + 2^2)(2^2 + 3^2) = 4^2 + 7^2 = 8^2 + 1^2.$$

Using Euclid's formulas for the general solution of $x^2 + y^2 = z^2$ (i.e., $z = p^2 + q^2$; $x = p^2 - q^2$; $y = 2pq$) we obtain two more right triangles with hypotenuse 65: $33, 56, 65$ and $63, 16, 65$. This completes the solution of the problem.

In connection with this problem Fermat stated that a prime of the form $4n + 1$ could be written as a sum of squares in just one way. Then he gave a formula for the determination of the number of ways in which a given number can be written as a sum of squares. Thus problems involving indeterminate equations led to number-theoretic insights.

Did Diophantus know the theorems formulated by Fermat? It is possible that he did. Jacobi offered a reconstruction of Diophantus' conjectured proofs, but the answer to this question remains hypothetical.

One can hardly overestimate the significance of Diophantus' *Arithmetica* for the subsequent history of algebra. It is no exaggeration to say that its role was comparable to the role of Archimedes' treatises in the history of the differential and integral calculus. We will see that it was the starting point for all mathematicians up to Bombelli and Viète, and that its importance for number theory and for indeterminate equations can be traced up to the present.

4. Algebra after Diophantus

The period from the 4th to the 6th centuries CE was marked by the precipitous decline of ancient society and learning. But eminent commentators, such as Theon of Alexandria (second half of the 4th century) and his daughter Hypatia (murdered in 418 by a fanatical Christian mob), were still active. In the 5th century there was an exodus of scholars from Alexandria to Athens. Finally, in the 6th century, Eutocius and Simplicius, the last of the great commentators, were expelled from Athens and settled in Persia.

We can turn to the question of the Arabic translations of four books attributed to Diophantus. An analysis of these books, translated at the end of the 9th century from Greek to Arabic by Costa ibn Luca (i.e., the Greek Constantin, son of Luca) shows that it is a reworked version of Diophantus' *Arithmetica*. It contains problems, possibly due to Diophantus, as well as extensive additions and commentaries to them. According to Suidas' Byzantine dictionary, Hypatia wrote commentaries on *Arithmetica*. It is therefore very likely that the four books translated into Arabic are books edited and provided with commentaries by Hypatia. These books contain no new methods, but the material is presented in a complete and systematic manner. Their author went beyond Diophantus by introducing the 8th and 9th powers of the unknown.

The subsequent development of mathematics, including that of algebra, was connected with the Arabic East. Scholars from Syria, Egypt, Persia, and other regions conquered by the Arabs wrote scientific treatises in Arabic. We will deal with their work in the next chapter.

Editor's notes

[1] For an extensive discussion of whether Diophantus realized that some of his problems had infinitely many solutions see pp. 28–29 in [1].

[2] In connection with the substitutions (6) and (11) the authors point out that Euler used such substitutions to rationalize the integrand in

$$\int \frac{dx}{\sqrt{ax^2 + bx + c}}.$$

Actually, Euler rationalized the integrand in

$$\int R(x, \sqrt{ax^2 + bx + c})\, dx,$$

where $R(x, y)$ is a rational function. Specifically, he used the substitution $y = \sqrt{a}x + t$ in the case $a > 0$ (cf. Diophantus' substitution (11)), the substitution $y = tx + \sqrt{c}$ in the case $c > 0$ (cf. Diophantus' substitution (6)), and a third substitution in the case when $ax^2 + bx + c$ had two different real roots.

CHAPTER **4**

Algebra in the Middle Ages in the Arabic East and in Europe

1. The emergence of algebra as an independent discipline

The fall of ancient society was accompanied by the decline of its science and culture. Science was destined to flourish again primarily in the Near and Middle East, in countries with a very old culture, and later in Western Europe.

In the 7th century CE a new religion—Islam—arose in Arabia, and the Arabs embarked on a series of conquests. In a short time they overran Persia, the states of Central Asia, Egypt, Western Africa, and a part of Spain. Soon a huge empire came into being which in its heyday extended from India to Spain. Many of its subject nations were culturally superior to the conquerors. These were the inhabitants of ancient Khorezm (a state in central Asia), as well as Persians, Syrians, Egyptians, and many others.

In the 8th century caliph Haroun al-Rashid (of *Arabian Nights* fame) established in the capital Baghdad of the new empire a "House of Wisdom", a variant of the Alexandrian Museon, which flourished between 819 and 833 under caliph al-Mamun. It had an extensive library. Here worked numerous scholars and translators. Arabic became the language of science throughout the lands of Islam. The Greek works translated into Arabic in the 9th century included Euclid's *Elements*, Ptolemy's *Almagest*, a number of works of Plato and Aristotle (referred to as the Great Philosopher), and (at the end of the 9th century) some of the books of Diophantus' *Arithmetica*.

Works of Indian scholars, so-called *Siddhāntas*, were also translated into Arabic. These works were mainly devoted to astronomy but sometimes included chapters on arithmetic and algebra.

The first discipline to flourish was astronomy and, in connection with astronomy, plane and spherical trigonometry and computational methods. These

49

interests were pursued by scholars of the medieval East for five to six centuries and they promoted the development of algebra.

The first eminent 9th-century scholar associated with the House of Wisdom was Mohammed ibn-Musa al-Khwarizmi. He was an encyclopedist, and his works dealt with mathematics, astronomy, cartography, and history.

Al-Khwarizmi was the author of two outstanding treatises devoted to arithmetic and algebra respectively. In the first of these he presented the decimal system of numeration, which came to the Arabic East from India, and its rules of operation. This work has come down to us in a Latin version that opens with the words "Dixit Algorithmi". Algorithmi is the Latinized form of al-Khwarizmi's name. The scholars of medieval Europe who adopted the decimal positional system of numeration referred to themselves as "algorithmists". This distorted version of al-Khwarizmi's name came to denote the computational rules associated with the new system of numeration. Later, the same word also referred to a method for solving each of a class of problems in a finite number of steps. This is how the name of this great Central Asian scholar was perpetuated.

The preserved version of the treatise on algebra is in Arabic. It is titled *A short book on the calculus al-jabr and al-muqābalah*. *Al-jabr* means "completion", i.e., the operation of adding a term to both sides of an equation, and *al-muqābalah* means "contrapositive", i.e., reduction of similar terms. These are the two operations referred to much earlier by Diophantus.

Al-Khwarizmi did not use algebraic symbolism but employed consistently three words for three powers of the unknown: 1) "dirhem" for number (i.e., for x^0); 2) "jizr" (root) or "shai" (thing) for an unknown x; and 3) "mal" (property, sum of money, also square) for x^2.

The treatise can be said to deal mainly with the solution of quadratic equations. What is remarkable is that Al-Khwarizmi classifies them; in other words, the equations are not regarded as a means for the solution of problems but as an independent object of study. He singles out the following six classes of quadratic equations:

1. "Squares equal roots", i.e., $ax^2 = bx$.
2. "Squares equal numbers", i.e., $ax^2 = c$.
3. "Roots equal numbers", i.e., $bx = c$.
4. "Squares and roots equal numbers", i.e., $ax^2 + bx = c$.
5. "Squares and numbers equal roots", i.e., $ax^2 + c = bx$.
6. "Roots and numbers equal squares", i.e., $bx + c = ax^2$, $a, b, c > 0$.

The solution involved three steps. The first was completion to a square. Thus in case of the equation

$$x^2 + 10x = 39$$

Al-Khwarizmi added to both sides 25 and obtained

$$(x + 5)^2 = 64.$$

The next step was extraction of roots; in the present case it yielded

$$x + 5 = 8, \quad x = 3.$$

The last step was a geometric justification of the solution which consisted in showing that the method was applicable to every equation of the form

$$x^2 + ax = b \quad a, b > 0.$$

The justification was very different from the one in Euclid's *Elements*. The details follow: the unknown x was represented by a segment, x^2 by the square on this segment, and $10x$ by two rectangles with sides x and 5 (Figure 16). The resulting gnomon was completed to a square by the addition of a square of side 5. Then the side of the resulting square was $x + 5$ as well as 8 (for the area of the large square was equal to $39 + 25 = 64$), whence $x = 3$.

Al-Khwarizmi's construction is an exact geometric copy of our algebraic procedure of "completing the square". In other words, the sole reason why Al-Khwarizmi resorted to a geometric interpretation for proving the generality of his method was that he had no algebraic symbolism for doing it in a non-geometric manner.

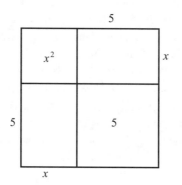

FIGURE 16

In the subsequent works of mathematicians of the medieval East we find a growing tendency to adopt the mathematical tradition of antiquity.

Four books attributed to Diophantus were translated into Arabic at the end of the 9th century (see Ch. 3). It is very likely that (possibly) edited versions of the first three books of *Arithmetica* were translated even earlier. Diophantus' methods had a decisive effect on the scientific activities of Arab mathematicians of the 10th and 11th centuries.

We have splendid treatises on algebra by Abu Kamil (ca. 850–ca. 930) and al-Karaji (d. 1016). Abu Kamil was also known under the name of al-Hasib al Misri, i.e., the Egyptian calculator. He came from Egypt and was evidently well acquainted with the tradition that went back to Diophantus. In his *Book of Algebra and Almucabala* he writes that the reader must know the three types of magnitudes mentioned by al-Khwarizmi (i.e., x^0, x^1, x^2) and adds to these five more powers: cube (kab)—x^3, square-square (mal al-mal)—x^4, square-cube (more correctly, square-square-thing, or mal mal shai)—x^5, cubo-cube (kab al kab)—x^6, and square-square-square-square (mal mal mal al-mal)—x^8. We see that his powers, like those of Diophantus, are subject to the additive principle and that he introduces a new power, namely x^8.

Part III of his treatise is devoted to the study of indeterminate equations. The study is marked by great depth and clarity (see [2]). We mention just his criterion for the solvability of the equation

$$y^2 = -x^2 + ax + b \tag{1}$$

over the rationals. By rewriting it as

$$y^2 + (x - a/2)^2 = b + (a/2)^2,$$

Abu Kamil arrives at the conclusion that the equation is solvable if and only if $b + (a/2)^2$, or equivalently $4b + a^2$, is representable as a sum of two squares.

The famous Persian mathematician al-Karaji brought to completion a phase of the development of indeterminate analysis and of the algebraic issues connected with it. In his treatise *Al-Fakhri* he transformed Diophantus' short algebraic introduction to his *Arithmetica* into a large treatise on algebra in which he introduced infinitely many positive and infinitely many negative powers of the unknown. He did this by means of what could be called two continued proportions. Using modern symbolism we could write them as follows:

$$x : x^2 = x^2 : x^3 = \cdots;$$
$$1/x : 1/x^2 = 1/x^2 : 1/x^3 = \cdots.$$

Al-Karaji employed words rather than symbols. For example, he described the second of his continued proportions in these words: "Know that the ratio of part of a thing $(1/x)$ to part of a square $(1/x^2)$, is like the ratio of part of a square $(1/x^2)$ to part of a cube $(1/x^3)$, is like the ratio of part of a cube to part of a square-square $(1/x^4)$, is like the ratio of part of a square-square to part of a square-cube $(1/x^5)$, and therefore the proportion of parts continues to infinity in accordance with this rule."

In this connection Al-Karaji formulated the following general rules:

$$\frac{1}{x^m} : \frac{1}{x^n} = \frac{x^n}{x^m}; \quad \frac{1}{x^m}\frac{1}{x^n} = \frac{1}{x^{m+n}}; \quad \frac{1}{x^m}x^n = x^{n-m}, \qquad n > m.$$

Then he introduced negative numbers and, like Diophantus, formulated for them the following multiplication rule: "Multiplying excess (zand) by excess we obtain excess, multiplying shortage (nakis) by shortage we obtain excess. In other cases we obtain shortage." In our notation this translates into the rules:

$$(+) \cdot (+) = (+); \quad (+) \cdot (-) = (-);$$
$$(-) \cdot (-) = (+); \quad (-) \cdot (+) = (-).$$

When explaining operations with polynomials he introduced rules which we would write as follows:

$$ax^m - bx^m = (a - b)x^m, \quad \text{if} \quad a > b;$$
$$ax^m - bx^m = -(b - a)x^m, \quad \text{if} \quad a < b;$$
$$ax^m - (-bx^m) = (a + b)x^m.$$

It is interesting to note that al-Karaji offered two justifications of the rule for the solution of quadratic equations, one geometric and the other according to "the method of Diophantus". The latter is purely algebraic. It involves the formation of a complete square followed by the extraction of a square root. Obviously, what counts is that al-Karaji realized that one can justify the rule of solution of quadratic equations algebraically, without recourse to geometry.

We mentioned that in his treatise abu-Kamil devoted a great deal of space to problems reducible to indeterminate equations. The same can be said of al-Karaji. The latter described algebra as the art of solving determinate and indeterminate equations. Almost all of the indeterminate equations handled by al-Karaji are taken from Diophantus' *Arithmetica* and are solved by his methods. Thus *Al-Fakhri* contains almost all of the problems in Books II and III of *Arithmetica* and all of the problems in Book IV of the Arabic version of this treatise. Many of al-Karaji's problems are borrowed from the work of abu-Kamil.

When dealing with equations of the form

$$y^2 = ax^2 + bx + c$$

al-Karaji notes that they are solvable if $a = \alpha^2$ or $c = \beta^2$. He also adduces abu-Kamil's criterion for the solvability of equations of the form

$$y^2 = -x^2 + bx + c.$$

In summary, we can say that in his treatise al-Karaji extended the domain of numbers and introduced arbitrary—positive and negative—integral powers of the unknown. Also, both he and abu-Kamil were profoundly influenced by Diophantus and his school.

We note that already abu-Kamil made extensive use in his treatise of algebraic irrationalities and applied to them transformations such as the following:

$$\sqrt{a} \pm \sqrt{b} = \sqrt{a + b \pm 2\sqrt{ab}}.$$

Also, he regarded them as arithmetical objects. This approach was continued by other mathematicians of the Arab East, who translated many of the geometrically defined irrational expressions in Book X of Euclid's *Elements* into the language of arithmetic. Thus in a work of al-Baghdadi (11th century) we find the following examples:

$$\sqrt{6 \pm \sqrt{20}} = \sqrt{5} \pm 1;$$

$$\sqrt{10} \pm \sqrt{8} = \sqrt{18 \pm \sqrt{320}}.$$

Between the 11th and 15th centuries algebraists gradually abandoned the Diophantine tradition. For example, in his treatise *On Proofs of Problems of Algebra and Almucabala* (*Historico-Mathematical Investigations*, Moscow, Gostekhteoretizdat, 1958, Issue 6. (Russian)), the eminent 11th-century poet and fine mathematician Omar Khayam used the geometric algebra of the ancients, and thus returned to the tradition of Euclid, Archimedes, and Apollonius. Following the classification principle applied by al-Khwarizmi to quadratic equations, Omar Khayam proposed a division of cubic equations into 27 classes, and instead of trying to express their roots in terms of radicals he tried, like Archimedes, to express them by using conic sections. But his analysis is nowhere near the sophisticated analysis of Archimedes; in fact, his treatise contains errors.

By the time of Khayam the mathematicians of the Arab East realized that the solution of (determinate and indeterminate) equations is a distinct discipline. Omar Khayam wrote: "The art of algebra and almucabala is a scientific art whose subject is absolute number and measurable magnitudes that are unknown but refer to some known thing that makes it possible to determine them. ... The aim of this art is to find relations connecting this subject with the specified data. The perfection of this art consists in knowing methods of investigation that enable one to perceive a way of determining the mentioned unknowns, numerical as well as geometric" (ibidem, p. 17).

Later, mathematicians of the East who worked in algebra in connection with astronomical investigations were primarily interested in numerical methods of solution of equations. This direction reached its highest level of development in the works of al-Kashi, who worked in the 15th century in Samarkand in the observatory of Ulugh Beg. In his encyclopedic treatise *Key to Arithmetic* al-Kashi used not only sexagesimal but also decimal fractions. In his treatise *On the Circumference* he inscribed in and circumscribed about a circle regular polygons with $3 \cdot 2^n$, $n = 1, 2, \ldots, 28$, sides and in this way computed π to 16 decimal places. Of course, in so doing he was applying the method first introduced by Archimedes. Finally, in the treatise *On the Chord and the Sine* he considered the problem of trisection of an angle, which reduces to the equation $\sin 3\varphi = 3 \sin \varphi - 4 \sin^3 \varphi$, or $x^3 + q = px$. He solved the latter by a brilliant iteration method. We note also that, beginning with al-Karaji, Arab mathematicians took an interest in the binomial formula $(a+b)^n$, $n \in \mathbf{Z}^+$. They used it to compute roots of arbitrary positive degree n.

In summary, in the Arab East algebra became an independent subject that dealt with the solution of determinate and indeterminate equations. In particular, arbitrary integral powers of the unknown and rules for operating with them were introduced.

Compared with the period of Diophantus, the one backward step was the failure to use literal symbolism. The unknown and its powers (and sometimes even numbers) were written down in words and this made algebra clumsy and hard to operate with.

2. The first advances in algebra in Europe

The first European achievements in algebra date back to the 13th century, the period known as the "early Renaissance". The algebraic tradition was transmitted by three routes: from the Arab East, from the Arabs who conquered Spain and established there the first advanced schools, and from Byzantium, which preserved the traditions of antiquity. The first major European mathematician was Leonardo of Pisa, or Fibonacci (ca. 1180–1240). He was born in Pisa, a commercial city state with large colonies in northern Africa stretching from Bugia (now in Algeria) to Sfaks (now in Tunisia).

Leonardo's father was a notary of the republic of Pisa. Shortly after Leonardo's birth his father was sent to Bugia on official business. His function there was similar to that of today's consul. When the boy was 12 his father sent for him. He was to learn about commerce and about arithmetical procedures. All this information is found in Leonardo's introduction to his fundamental *Book of the Abacus* (Liber abaci, Scritti di Leonardo Pisano, Roma, 1862). He traveled to Egypt, Syria, and Provence and familiarized himself with different methods of calculation as well as with the rudiments of algebra. He concluded that for purposes of calculation the decimal positional system (which he referred to as Indian) was far superior to all other systems. Upon his return to Pisa he began to study mathematics in earnest. In particular, he studied Euclid's *Elements*. By combining this knowledge with what he learned from the Arabs, Leonardo wrote his famous *Book of the Abacus*, unequaled for 300 years. It contained information about arithmetic, algebra, and geometry. It was not just an anthology. Many of its problems were dealt with in an entirely original way. For example, in connection with his "rabbit problem" (How many pairs of rabbits will be produced in a year, beginning with a single pair, if in every month each pair bears a new pair which becomes productive from the second month on?) he introduced the famous Fibonacci series

$$1 + 1 + 2 + 3 + 5 + 8 + 13 + \cdots, \tag{2}$$

now used extensively in biology for the description of a large variety of processes as well as in computational mathematics (there is a number system for computers based on the entries in this series). The Fibonacci series is recursive (a series is recursive if each of its entries is a linear combination of some of its predecessors), for

$$u_{n+1} = u_n + u_{n-1}.$$

In addition to the *Book of the Abacus* Leonardo wrote two extremely interesting books. They are *Flower* (Flos, 1225) and *Book of Squares* (Liber quadratorum, 1225). In these books he investigated problems he was challenged with by magister Johann of Palermo, court philosopher of Frederic II, king of Sicily. The first problem was to solve the cubic equation

$$x^3 + 2x^2 + 10x = 20,$$

and the second was to find a rational solution of the system

$$\begin{cases} x^2 + 5 = u^2, \\ x^2 - 5 = v^2, \end{cases} \tag{3}$$

which is a special case of the system

$$\begin{cases} x^2 + a = u^2, \\ x^2 - a = v^2. \end{cases} \tag{3'}$$

To solve the first problem Leonardo analyzed it in great detail. He rewrote it as

$$\frac{x^3}{10} + \frac{x^2}{5} + x = 2$$

and showed that $1 < x < 2$, i.e., that the positive root cannot be an integer. Then he showed that it cannot be a fraction. His proof is entirely general. It can be used to show that if the equation

$$x^n + a_1 x^{n-1} + \cdots + a_{n-1} x + a_n = 0,$$

$a_i \in \mathbf{Z}$, has no integral root then it has no rational root. This is an important theorem of algebraic number theory first proved in the 19th century.

We reproduce Leonardo's argument. Suppose the equation had a rational root $x = p/q$, $(p, q) = 1$. Then

$$p^3 + 2p^2 q + 10pq^2 = 20q^3.$$

Since all terms of this equation except the first are divisible by q, p must also be divisible by q. Contradiction.

Leonardo went on to show that the root cannot be of the form $\sqrt{p}$, $p \neq \alpha^2$. Indeed, the root x can be written as

$$x = 2\frac{10 - x^2}{10 + x^2}.$$

If $x = \sqrt{p}$, then we would have $\sqrt{p} = 2\frac{10-p}{10+p}$, i.e., $\sqrt{p}$ would be rational. But this contradicts the previous insight.

Finally, he showed that the root cannot be represented by any of the irrationalities in Book X of Euclid's *Elements* and gave its approximate sexagesimal value as

$$x = 1^0 22' 7'' 43''' 33^{\mathrm{IV}} 4^{\mathrm{V}} 40^{\mathrm{VI}}.$$

No analysis of comparable depth of this problem can be found until the end of the 18th century.

Leonardo considered the second problem in his *Book of Squares*. First he solved the following two problems in Book II of Diophantus' *Arithmetica*:

$$x^2 + y^2 = a^2; \tag{4}$$

$$x^2 + y^2 = a^2 + b^2 \neq \square. \tag{5}$$

Leonardo used a different method to solve each of these problems. Apparently, by that time some of the statements of Diophantus' problems had reached Europe but not his methods of their solution.

To solve problem (4) Leonardo takes a right triangle ABC with rational sides (p, q, r) (Figure 17) and lays off on its hypotenuse r (or on its extension) a segment AM of length a. From M he drops the perpendicular MN to the side AC. Then $x = AN = p(a/r)$ and $y = MN = q(a/r)$ are solutions.

Before solving problem (5) he proves the formula for composition of forms

$$(a^2 + b^2)(p^2 + q^2) = (ap - bq)^2 + (aq + pb)^2 = (ap + bq)^2 + (aq - pb)^2. \tag{6}$$

We recall that this formula was known to the Babylonians and was used by Diophantus in his *Arithmetica*. Leonardo notes that if $a/p \neq b/q$, then formula (6) yields two different representations for the product of forms; if

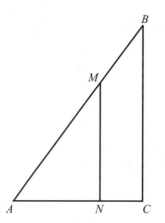

FIGURE 17

$a/p = b/q$, then we obtain just one representation:

$$k^2(a^2 + b^2)^2 = k^2(a^2 - b^2)^2 + k^2(2ab)^2. \tag{6'}$$

To solve problem (5) he takes a right triangle with rational sides (p, q, r) and carries out the composition of forms (6). Then he takes a right triangle with legs $|ap - bq|$ and $aq + bp$ and hypotenuse $r\sqrt{a^2 + b^2}$ and a similar triangle with legs $|ap - bq|/r$ and $(aq + bp)/r$ and hypotenuse $\sqrt{a^2 + b^2}$. In this way he obtains the following solution of equation (5):

$$x = \frac{|ap - bq|}{r}; \quad y = \frac{aq + bp}{r}$$

$(a^2 = 16, \ b^2 = 25, \ p = 3, \ q = 4, \ r = 5, \ x = 8/5, \ y = 31/5)$.

A second solution of this equation is

$$x_1 = \frac{ap + bq}{r} \quad \left[= \frac{32}{5}\right]; \qquad y_1 = \frac{|aq - bp|}{r} \quad \left[= \frac{1}{5}\right].$$

Later Viète used this unusual solution (see Ch. 5).

Now Leonardo turns to the solution of the system (3), or, equivalently, of $(3')$ for $a = 5$, and finds that $x = 41/12$. He notes, furthermore, that for $(3')$ to be solvable over the integers, with x, u, v a triple of pairwise coprime integers, a must be of the form $4\,kl(k + l)(k - l)$. He calls this expression a congruum. It is easy to see that the congruum is four times the area of the right triangle with legs $k^2 - l^2$ and $2kl$. Leonardo claims that a congruum cannot be a square. This is equivalent to the assertion that the area of a right triangle with rational sides cannot be a square. The latter claim implies Fermat's Last Theorem for $n = 4$, i.e., the unsolvability of the equation

$$x^4 + y^4 = z^4$$

over the (positive) integers. Thus Leonardo formulated Fermat's Last Theorem for the case $n = 4$ four hundred years before Fermat. His proposed proof contains an error.

3. Algebraic symbolism in Europe. The German cossists and the development of algebra in Italy

For almost 300 years there was neither a European scholar of Leonardo's rank nor a European scholar capable of understanding and appreciating the riches contained in his works. It was not until the second half of the 15th century, the century of the Renaissance, that we witness a revival of algebraic

04

investigations. This revival was aided by two world-scale events: 1) the fall of Constantinople (Byzantium) in 1453 and the migration of Greek scholars to Europe and 2) the invention of the printing press.

It was in Byzantium that ancient Greek manuscripts were for a long time preserved, copied, and commented on. The famous French historian of science Paul Tannery found a letter by the 11th-century Byzantine mathematician Michael Psellus (included in his critical edition of the works of Diophantus published in 1893) which showed that Byzantine mathematicians of that time knew three ways of denoting an unknown and its powers.

1. Notation based on the additive principle, of the kind used by Diophantus. There are three special symbols for the first three powers of the unknown and the remaining powers are composed from these three using the principle of addition of exponents: $x^4 = \Delta^v\Delta$; $x^5 = \Delta K^v$; $x^6 = K^v K$ (see Ch. 3, 2).

2. Notation based on the multiplicative principle. This system goes back to Anatolius of Laodicea, author of an *Introduction to Arithmetic* and a contemporary of Diophantus. The names of the first three powers and the symbols for them were the same as in the case of Diophantus:

x	$\overset{\prime}{\alpha}\rho\iota\theta\mu\acute{o}\varsigma$ or $\pi\lambda\epsilon\upsilon\rho\acute{\alpha}$	(side)
x^2	$\delta\overset{\prime\prime}{\upsilon}\nu\alpha\mu\iota\varsigma$	(square)
x^3	$\varkappa\overset{\prime}{\upsilon}\beta o\varsigma$	(cube)

The names of higher powers were based on the multiplicative principle:

x^4	$\delta\overset{\prime}{\upsilon}\nu\alpha\mu o\delta\acute{\upsilon}\nu\alpha\mu\iota\varsigma$	(square–square)
x^5	$\overset{\prime}{\alpha}\lambda o\gamma o\varsigma \pi\rho\tilde{\omega}\tau o\varsigma$	(first inexpressible)
x^6	$\delta\overset{\prime\prime}{\upsilon}\nu\alpha\mu o\varkappa\acute{\upsilon}\beta o\varsigma$	(square–cube)
x^7	$\overset{\prime}{\alpha}\lambda o\gamma o\varsigma \delta\epsilon\acute{\upsilon}\tau\epsilon\rho o\varsigma$	(second inexpressible), and so on.

Note that the second inexpressible, i.e., x^7, is the product of the first inexpressible by x^2. In this notational system $\varkappa\upsilon\beta o\varkappa\upsilon\beta o\varsigma$ (cube-cube) stands for x^9 and not, as in Diophantus' case, for x^6.

3. A notational system attributed to Michael Psellus. He named the powers of the unknown simply in terms of their order:

x	$\overset{\prime}{\alpha}\rho\iota\theta\mu\acute{o}\varsigma \pi\rho\tilde{\omega}\tau o\varsigma$	(first number)
x^2	$\overset{\cdot}{\alpha} \delta\epsilon\acute{\upsilon}\tau\epsilon\rho o\varsigma$	(second number)
x^3	$\overset{\cdot}{\alpha} \tau\rho\iota\tau o\varsigma$	(third number)
x^4	$\overset{\cdot}{\alpha} \tau\epsilon\tau\alpha\rho\tau o\varsigma$	(fourth number)
x^5	$\overset{\cdot}{\alpha} \pi\epsilon\mu\pi\tau o\varsigma$	(fifth number), and so on.

It is interesting that the notational system that became popular in Europe in the 15th century was a system based on the multiplicative principle. We regard such a system as rather inconvenient.

We saw that the Arab algebraists did not use symbols for the unknown and its powers, so that the idea of literal notation could not come to Europe from the Arab East. Leonardo Pisano used segments (like Euclid) or special terms, such as radix (root), res (thing), and census (square, property) for denoting an unknown. These words were translations of the corresponding Arab terms. He also used the term censo di censo (square-square) for the fourth power of the unknown.

The mathematician Jordanus Nemorarius is important in the present context. He was a contemporary of Leonardo but definitely not of his professional stature. In his *Arithmetic Presented in Ten Books* he systematically used letters for concrete numbers. He used neither segments nor rectangles for expressing magnitudes, so that with him literal notations are pure symbols for arbitrary numbers. He invariably expressed a magnitude by a single letter. Since he did not use symbols for equality and for algebraic operations, he was forced to introduce new literal notations for every new result. It is this that prevented him from developing a literal calculus.

In the 16th century Maurolico (Francesco de Messina) proceeded in the same way, and so his notational system was marred by the same flaw.

Beginning in the 12th century Arab mathematicians in the West began to use for the unknown and its powers the first letters of words rather than whole words. The Latin translations of the Arab words were cosa, censo, and cubus. This forced Italian mathematicians to use the first two letters of each of these words. Thus in his *Questions of Algebra* (Questioni d'Algebra, 1384) Maestro Gilio used co. (from cosa) for x, ce. (from censo) for x^2, cu. (from cubo) for x^3, and ce.ce. (from Leonardo's censo di censo, square-square) for x^4. Gilio used R, the first letter of radice, to denote a square root and circled it if another square root appeared under the radical sign (another tradition derived from Leonardo). For addition and subtraction he used the words piu (plus) and meno (minus). Thus Gilio would write $5 - \sqrt{25 - \sqrt{10}}$ as 5 meno Ⓡ di 25 meno R di 10.

In 1494 Luca Pacioli published his *Summary of the Knowledge of Arithmetic, Geometry, Proportions and Proportionalities* (Summa de arithmetica, geometria, proportioni et proportionalita). This was one of the first printed books. It contains a table (see Table 1 below) that can safely be called the fulfillment of efforts undertaken in Italy in the 14th and 15th centuries to perfect a system of notation for the unknown and its powers. Pacioli was the first to introduce symbols for the first 29 powers of the unknown.

Pacioli uses the symbol n° (short for numero) for the constant term in an equation. The word relato is a translation of the Greek $\overset{\prime}{\alpha}\lambda o\gamma o\varsigma$ (inexpressible) probably misread as $\acute{o}\lambda o\gamma o\varsigma$ (relation). We see that the table is based on the multiplicative principle and the first 29 powers of the unknown include 9 "inexpressibles"!

Pacioli's book also contains symbols for a second unknown y, called quantità (quantity) and denoted by q. p°, as well as for its square y^2, denoted by ce. di q. p°.

Table 1
NOTATIONS FOR THE UNKNOWN AND ITS POWERS

1. Diophantus of Alexandria (*3rd century* CE)

x°	$\overset{\circ}{\mathrm{M}}$	$\mu\acute{o}\nu\alpha\varsigma$	unity
x^1	ς	$\overset{\prime}{\alpha}\rho\iota\theta\mu\acute{o}\varsigma$	number
x^2	Δ^υ	$\delta\overset{\prime}{\upsilon}\nu\alpha\mu\iota\varsigma$	power, degree
x^3	K^υ	$\kappa\acute{\upsilon}\beta o\varsigma$	cube
x^4	$\Delta^\upsilon\Delta$	$\delta\overset{\prime}{\upsilon}\nu\alpha\mu o\delta\acute{\upsilon}\nu\alpha\mu\iota\varsigma$	square–square
x^5	$\Delta\,\mathrm{K}^\upsilon$	$\delta\overset{\prime}{\upsilon}\nu\alpha\mu o\kappa\acute{\upsilon}\beta o\varsigma$	square-cube
x^6	$\mathrm{K}^\upsilon\,\mathrm{K}$	$\kappa\overset{\prime}{\upsilon}\beta o\kappa\acute{\upsilon}\beta o\varsigma$	cube-cube

2. Adam Ries (*1489–1559*)

x°	$\varnothing$	Dragma, Numeros	number
x^1	$\mathfrak{r}$	Radix, Coss	root, thing
x^2	$\mathfrak{z}$	Zensus	square
x^3	$\mathfrak{c}$	Cubus	cube
x^4	$\mathfrak{zz}$	Zensus de Zensu	square-square
x^5	β	Sursolidum	deaf solid
x^6	$\mathfrak{zc}$	Zensuicubus	square-cube
x^7	$bi\beta$	Bissursolidum	second deaf solid
x^8	$\mathfrak{zzz}$	Zensus Zensui de Zensu	square-square-square
x^9	$\mathfrak{cc}$	Cubus de Cubo	cube-cube

3. Luca Pacioli (*ca. 1445–ca. 1515*)

x°	n°	numero	number
x^1	co.	cosa	thing
x^2	ce.	censo	square

x^3	cu.	cubo	cube
x^4	ce.ce.	censo di censo	square-square
x^5	p°.r°.	primo relato	first inexpressible
x^6	ce.cu.	censo di cubo	square-cube
x^7	2°.r°.	secundo relato	second inexpressible
x^8	ce.ce.ce.	censo de censo de censo	square-square-square
x^9	cu.cu.	cubo de cubo	cube-cube
x^{10}	ce.p°.r°.	censo de primo relato	square of first inexpressible
x^{11}	3°.r°.	terco relato	third inexpressible
. . .	. . .	. . .	. . .
x^{29}	9°.r°.	nono relato	ninth inexpressible

4. Rafael Bombelli (ca. 1526–1573)

Bombelli used $\underset{\smile}{1}$ for the unknown and $\underset{\smile}{2}$, $\underset{\smile}{3}$,. . . for its powers.

5. Simon Stevin (1548–1620)

Stevin used the symbols ①,②, ③, . . . for one unknown, the symbols sec.①, sec.②, sec.③, . . . for a second unknown, the symbols ter.①, ter.②, ter.③, . . . for a third unknown, and so on.

Stevin also used symbols for various arithmetic operations: p̃ for addition (the tilde over p indicates the use of an abbreviation for the word più), m̃ for subtraction (the abbreviated word is meno), ℞. for a square root (the abbreviated word is radice), ℞.3. for a cube root, and ℞.4. (or ℞. ℞.) for a fourth root. These were the traditonal notations of the Italian school of abacists at the time.

Luca Pacioli introduced negative numbers in the manner of Diophantus (who was imitated earlier by al-Karaji), i.e., by an axiomatic definition of the rules of operation with the new numbers:

> Piu via piu sempre fa piu.
>
> Meno via meno sempre fa piu.
>
> Piu via meno sempre fa meno.
>
> Meno via piu sempre fa meno.

Here piu stands for a positive number and meno for a negative one.

In order to justify the reasonableness of his definition Pacioli gives examples which show that it leads to results which agree with the usual rules of arithmetic (we would say today that the results agree with the distributive laws), for example: (using modern notations) $8 \cdot 8 = 64 = (10 - 2)(10 - 2) =$

$100 - 20 - 20 + 4 = 64$, i.e., for the outcomes to agree we must put

$$\bar{m}.\bar{p}.=\bar{m}.,$$

$$\bar{m}.\bar{m}.=\bar{p}.,$$

as was done in the definition of the rules of operation cited above.

It is thought that the notational system of powers of the unknown, introduced by Diophantus and based on the additive principle, displaced the corresponding system based on the multiplicative principle only after European mathematicians became familiar with Diophantus' *Arithmetica* in the 16th century. Apparently this process began earlier. A relevant example is the following table of symbols from Dionisius Gori's *Book of Practical Algebra* (Libro et trattato della praticha d'alcibra), published in 1544, which is free of inexpressibles:

1. co.	6. ce.cu.	11. cu.cu. R. R	16. ce.ce.ce.ce.
2. ce.	7. ce.cu. R	12. cu.ce.ce.	17. ce.ce.ce.ce. R
3. cu.	8. ce.ce.ce.	13. cu.ce.ce. R	18. ce.cu.cu.
4. ce.ce.	9. cu.cu.	14. ce.ce.cu. R	19. ce.cu.cu. R
5. ce.ce. R	10. cu.cu. R	15. cu.ce.ce. R	20. ce.ce.ce.ce. R

It is not difficult to see that Gori uses in his notational system of powers of the unknown an additive-multiplicative principle: $x^5 = x^{2 \cdot 2+1}$, $x^7 = x^{2 \cdot 3+1}$, $x^{14} = x^{2 \cdot (2 \cdot 3+1)}$, $x^{15} = x^{3 \cdot (2 \cdot 2+1)}$, $x^{20} = x^{2 \cdot 2 \cdot (2 \cdot 2+1)}$. He seems unaware that his principle is nonunique: the symbols for x^{13} and x^{15} are the same, and so are the symbols for x^{17} and x^{20}.

We must also note the contribution of German mathematicians to the development of symbolism. We mentioned the 16th-century Italian schools of abacists. A similar school came into being in Germany. Its name was *Coss*. The word derives from the Latin cosa.

The first to lecture on algebra in Germany was Johannes Widman (ca. 1462–d. after 1498). He was Bohemian by birth. Our + and − signs appeared for the first time in his book *A Quick and Beautiful Method of Calculation for all Merchants* (Behend und hüpsch Rechnung uff allen Kauffmanschafften), published in 1489. Widman wrote: "− is the same as shortage and + is the same as excess." And in Christoff Rudolff's textbook of 1525, the first to be written in German, these symbols are widely used.

The most famous cossist was Adam Ries (1492–1559), who published many influential textbooks on arithmetic. In 1525, he wrote a *Coss* manuscript (not printed until 1892) which accurately reflected the state of algebra in his day.

The first cossist algebras employed an additive principle for powers of the unknown. A relevant example is a manuscript of ca. 1480 used by Widman. But a manuscript of 1550, used by Rudolff, employed a multiplicative principle. The same principle was used by Ries (see Table 1).

"Just a number" is denoted by $\varnothing$, a defective way of writing the letter d (with a flourish), the first letter in Dragma, derived from the word *drikhem*. The latter was used by Arab mathematicians to denote the constant term in a quadratic equation. The symbols $\mathfrak{r}$, $\mathfrak{z}$ and $\mathfrak{c}$ are Gothic versions of the first letters in res, zensus, and cubus respectively. Of sursolidum Ries said that it was a "deaf number" (eine taube Zahl). In the manuscript of the *Founder of Algebra*, x^5 is called surdum solidum (deaf solid). The German taub and the Latin surdum are translations of the Arabic asam, used by the Arabs for the Greek $\overset{\prime}{\alpha}\lambda o\gamma o\zeta$ (inexpressible). The use of the word solidum indicates that the cossists regarded certain powers as generalized cubes.

We note that a characteristic feature of German algebra was the tendency to reduce the number of cossist symbols and the clear realization of the need for uniform symbolic notations. In this connection we must mention Michael Stifel's (1487–1567) *Complete Arithmetic* (Arithmetica integra) of 1544 which, in a sense, brought to a close the evolution of algebraic symbolism. Here Stifel introduced our present symbol for a square root and denoted unknowns by capital Latin letters $A, B, C, \ldots$, repeated as many times as indicated by the degree of the unknown. Stifel's approach was more suitable for the creation of a literal calculus than were the approaches of, say, Jordanus Nemorarius or of Maurolico. Many of Stifel's notations were adopted not only in Germany but also in Italy.

An important contribution to the development of algebraic symbolism was made by Nicolas Chuquet, holder of the baccalaureate in medicine (d. ca. 1500). Chuquet wrote the exponent of the power of an unknown after the numerical coefficient; for example, he wrote $12x^3$ as 12^3. He also introduced into his symbolism the zeroth power of the unknown (for the constant term in an equation) as well as its negative powers. Thus he would write $8x^3 \cdot 7x^{-1} = 56x^2$ as "8^3 multiplied by $7^{1\cdot\tilde{m}}$ yields 56^2." Later Rafael Bombelli and Simon Stevin introduced similar notation for the unknown and its powers. This emphasized the homogeneity of the sequence of powers (see Table 1).

The algebraic part of Luca Pacioli's famous *Summa*. . ., in which he considered the solution of indeterminate equations (following Leonardo Pisano) and the solution of determinate equations of degrees one and two, ended with the statement that just as there is no method for effecting the quadrature of a circle so, too, there is no general method for the solution of cubic equations of the form $x^3 = ax + b$ and $x^3 + ax = b$. Pacioli wrote this the day before such a method was found.

The first great advances in mathematics in Europe were connected with the solution of algebraic equations of degrees three and four. There is no doubt that the introduction of literal symbolism played an important role in this development. From then on the evolution of algebra was given a new direction which led at the end of the 16th century to the creation of the first literal calculus and to a new extension of the domain of numbers.

CHAPTER 5

The first achievements of algebra in Europe

1. The solution of cubic and quartic equations

The Renaissance—the 15th and 16th centuries—was a period of flourishing of art, science, and literature in Italy, Spain, France, and England, as well as a period of familiarization with the heritage of antiquity. At the time this period also seemed to be one of renaissance of the high culture of Greece and Rome. Actually, the acquired knowledge of antiquity made possible the building of foundations for a new science and culture in many respects very different from the science and culture of antiquity. The obvious leader in this movement was Italy, which then boasted brilliant painters (Botticelli, da Vinci, Raphael), sculptors (Michelangelo, Cellini), and (somewhat earlier, in the 14th century) writers and poets (Boccaccio, Petrarch).

This was also the time of great geographical discoveries and achievements, such as Columbus' discovery of America (1492) and Magellan's circumnavigation of the globe (1521). Girolamo Cardano (1501–1576), a striking representative of the Renaissance, a medical doctor, a mathematician, and a philosopher, described the new discoveries in these words: "I was born in this century in which the whole world became known; whereas the ancients were familiar with but little more than a third part of it. . . .. For what is more amazing than pyrotechnics? Or the fiery bolts man has invented, so much more destructive than the lightning of the gods? Nor of thee, O Great Compass, will I be silent, for thou dost guide us over boundless seas, through gloomy nights, through the wild storms seafarers dread, and through the pathless wilderness. The fourth marvel is the invention of the typographic art, a work of man's hands, and a discovery of his wit—a rival, forsooth, of the wonders wrought by divine intelligence. What lack we yet unless it be the taking of Heaven by storm!" (G. Cardano, *The Book of my Life*, tr. J. Stoner, Dover, 1962.)

When it comes to mathematics, the 16th century was the age of algebra. It began with the solution of cubic and quartic equations by radicals, the first great achievement beyond those of antiquity, and ended with the construction of a literal calculus and the introduction of complex numbers.

The history of the solution of cubic equations resembles a detective story. It began when Scipione del Ferro (1456–1526), a professor at the very old university of Bologna (founded at the beginning of the 12th century), solved by radicals the equation

$$x^3 + px = q, \quad p, q > 0, \tag{1}$$

but kept his method and results secret. Keeping a method secret was then just as common as today's tendency to publish one's discoveries as quickly as possible. The owner of a method could challenge his rival to a scientific duel and set him problems solvable by the method the rival was ignorant of. Victory in such a "tournament" brought one fame and placed one at advantage when it came to filling a desirable position. Before his death del Ferro disclosed his method to his student Fiore. In 1535 this student, not very gifted mathematically, challenged to a duel the well known Italian mathematician Niccolò Tartaglia.

Niccolò Tartaglia (ca. 1499–1557) was born into a poor family in the town of Brescia. His father, who brought the mail to his home town, died when the boy was six. When the French sacked Brescia in 1512 the boy was seriously wounded in the jaw and larynx. His mother was too poor to consult a doctor and treated him with home remedies. He stammered for years. Actually Tartaglia is a nickname rather than a name and stands for "stammerer".

In spite of all these difficulties Niccolò managed to acquire on his own a great deal of knowledge of mathematics and mechanics. He subsequently wrote an impressive—for his time—treatise, titled *The New Science*, in which he considered a variety of mechanical issues, including the computation of trajectories of projectiles. He lectured for a long time in his home town as well as in Verona and in Venice. He was also invited as a consultant for drawing up commercial contracts.

Tartaglia tells us that he was aware that Fiore was in possession of the late del Ferro's rule for the solution of equation (1) and made strenuous efforts to discover it himself. He succeeded the night before the duel. The duel took place on February 12, 1535. All of the 30 problems set by Fiore were special cases of equation (1) and Tartaglia solved them without difficulty. On the other hand, Tartaglia picked problems from different areas of mathematics and mechanics and Fiore was unable to solve any of them.

In solving equation (1) Tartaglia assumed that one of its roots is of the form

$$x = u - v.$$

Then the equation can be reduced to the form

$$u^3 - v^3 + (u - v)(p - 3uv) = q.$$

If one imposes on u and v the additional condition $3uv = p$, then u and v can be determined from the system

$$\begin{cases} u^3 - v^3 = q, \\ uv = p/3. \end{cases}$$

Putting $z = u^3$ we see that this system is equivalent to the quadratic equation

$$z^2 - qz - (p/3)^3 = 0,$$

which means that

$$x = \sqrt[3]{\sqrt{\left(\frac{q}{2}\right)^2 + \left(\frac{p}{3}\right)^3} + \frac{q}{2}} - \sqrt[3]{\sqrt{\left(\frac{q}{2}\right)^2 + \left(\frac{p}{3}\right)^3} - \frac{q}{2}}.$$

A few days after the duel Tartaglia was able to solve the equation

$$x^3 = px + q, \quad p, q > 0, \tag{2}$$

by using the substitution $x = u + v$. The corresponding formula was

$$x = \sqrt[3]{\frac{q}{2} + \sqrt{\left(\frac{q}{2}\right)^2 - \left(\frac{p}{3}\right)^3}} - \sqrt[3]{\frac{q}{2} - \sqrt{\left(\frac{q}{2}\right)^2 - \left(\frac{p}{3}\right)^3}}, \tag{3}$$

and it involved fundamental difficulties. Indeed, if $(q/2)^2 < (p/3)^3$, then under the square root in formula (3) there is a negative number, i.e., the whole formula becomes meaningless. On the other hand, examples showed that in this case equation (2) could have real roots (in fact, this is precisely the case when all three roots are real!); for example, the equation

$$x^3 = 15x + 4$$

has the root $x = 4$ in spite of the fact that $(4/2)^2 < (15/3)^3$. Thus one could not stipulate that $(q/2)^2 \geq (p/3)^3$. No such difficulty arises in the case of quadratic equations. In other words, one was led to investigate square roots of negative numbers in connection with the formula for a root of a cubic rather than of a quadratic equation!

The case when $(q/2)^2 < (p/3)^3$ was called "irreducible", for in this case the expression (3) did not yield a real root. It seems that the "irreducible" case troubled Tartaglia a great deal. He delayed the publication of his results because he was trying to resolve the difficulty. But it was at that very time that fate brought him together with the mathematician Girolamo Cardano, a man of impetuous character who would stop at nothing to achieve an objective.

Cardano was a true Renaissance figure and embodied the good and bad characteristics of that period. He was born into a family of a notary and spent his childhood in Milan. Already as a youth he had an obsessive need for fame. He wrote: "The aim I pursued was not riches or idleness, not honors, not high positions, not power but, insofar as I could achieve it, the immortalization of my name. . .."

Cardano studied medicine and became famous as a skillful surgeon. In addition he studied mathematics, mechanics, astrology, philosophy, and questions of education. He was an excellent fencer and bragged that "in my youth I was anxious to contend with stronger ones" and that "I could, unarmed, knock out a bared dagger from the hand of my opponent." He traveled to Scotland and attended members of the English and Scottish aristocracy and was for some time the court physician of the young Edward VI. He cast the king's horoscope and predicted for him a long life. But Edward died a few months later and Cardano returned home. He was shattered with grief when his beloved older son was sentenced to death for poisoning his wife. For some time he was a professor at the university of Bologna. Following a charge of involvement with black magic, he was pursued by the inquisition, chased out of Bologna, and deprived of the right to lecture. He spent the last years of his life in Rome where he was invited as a famous physician by Pope Pius V. Here he resumed his medical practice and in the last year of his life wrote his famous autobiography *The Book of my Life*, which has been translated into all European languages.

There is a story to the effect that Cardano cast his own horoscope and predicted that he would die on September 21, 1576. To boost his fame as an astrologer he presumably starved himself so as to die at the predicted time. True or not, the story is an excellent reflection of the essence of Cardano's character.

Cardano was in the process of writing a book on algebra when he found out that Tartaglia knew the secret of the solution of the cubic equation. Thereupon he made every effort to ferret out the secret. He sent Tartaglia an invitation to Milan in the name of a famous signor who "happened" to be out of town when Tartaglia arrived there. Tartaglia accepted Cardano's hospitality and told him the secret of the solution when Cardano swore not to disclose it. This happened in 1539. But in 1545 Cardano published his *Great Art, or The Rules of Algebra* (Ars magna sive de Regulis algebraicis) which included the rules for the solution of equations (1) and (2) as well as the solution of the quartic equation, discovered by Cardano's student Luigi Ferrari (1526–1565). And while it is true that Cardano referred in the first chapter of his book to del Ferro and to Tartaglia as the discoverers of the solution formula, the fame has remained his; to this day the formulas for the solution of equations (1) and (2) bear his name.

Rather than pursue the dramatic story of the relations between Tartaglia and Cardano, in which Cardano's students also played a significant role, we will return to cubic equations.

Cardano, like Tartaglia, was baffled by the "irreducible case". When solving the problem of finding two numbers x and y such that

$$x + y = 10, \quad xy = 40,$$

he noted that the expressions $5 + \sqrt{-15}$ and $5 - \sqrt{-15}$ satisfied the two conditions provided that one put $\sqrt{-15} \cdot \sqrt{-15} = -15$. However, he did not try to use expressions of the form $\sqrt{m}$, $m < 0$, for dealing with the irreducible case.

2. The *Algebra* of Rafael Bombelli. Introduction of complex numbers

We know as little about Rafael Bombelli (ca. 1526–1573) as we do about Diophantus, except that the dates of his birth and death are more certain. We do know that he lived in Bologna and was an accomplished hydraulic engineer. His *Algebra* shows that he was one of the most eminent algebraists of the modern era. Three parts of his book were published in 1572, but the remaining two, which contain methods in many ways superior to those of Descartes, were published only in 1929.

An early version of the manuscript, very different from the final one, was ready in 1550 but Bombelli kept on modifying it. Then the Roman mathematician Pacci informed him that he had found in the Vatican library the manuscript of a "certain Diophantus", a Greek author, devoted to arithmetic and algebra. Bombelli familiarized himself with the manuscript and was tremendously impressed by its contents. He decided to translate it "in order to enrich the world with such a remarkable work". He did not manage to complete the translation of the work but was so influenced by it that he reworked his own manuscript in a fundamental way.

The changes were especially marked when it came to: the domain of numbers, the manner of introducing complex numbers, the manner of introducing powers of the unknown, and the more abstract style of presentation. Whereas in the earlier version of the manuscript problems were stated in the traditional "amusing" or "pseudopractical" form, in the new version they were stated abstractly using abstract numbers.

A change of a very special nature was the inclusion, in Part III of the reworked version, of 143 problems with solutions taken from Diophantus' *Arithmetica*. It was from Bombelli's *Algebra* that European mathematicians first found out about these problems and the methods of their solution.

We now give a more detailed description of the contents of the *Algebra* (L'Algebra) of 1572. The first part is devoted to the construction of the domain of numbers needed for the development of algebra and to the introduction of powers of the unknown. First Bombelli introduces successive integral powers of rational numbers. He uses the multiplicative principle, which is why he calls the 5th power primo relato ("first inexpressible") and the 6th power square–cube. This is a concession on his part to a tradition that developed in European mathematics between the 14th and 16th centuries. There are very few traces of this tradition in Bombelli's book. As a rule, he names powers according to their exponents: "fifth", "sixth", and so on.

Next Bombelli introduces irrational magnitudes—square roots denoted in the manner of Pacioli as R.q., cube roots denoted as R.c., square–square roots denoted as R.R.q. and so on, as well as binomials and trinomials composed of these irrationalities. He considers arithmetical operations on all these magnitudes and sometimes resorts to geometric proofs.

Then Bombelli introduces meno, a negative number, in exactly the same way as Diophantus, i.e., by defining a "rule of signs" under multiplication. It takes the form of a table in which piu stands for plus and meno for minus:

> piu via piu fa piu
> meno via piu fa meno
> piu via meno fa meno
> meno via meno fa piu
> (R. Bombelli, *L'Algebra*, Milano, 1966, p. 62).

Like Pacioli before him, Bombelli adds explanations to his axiomatic definition. Specifically, he shows that if we want to preserve the distributive property of multiplication over addition, then we must stipulate that $(-) \cdot (-) = (+)$.

Bombelli notes that the same law of signs holds for division. He also formulates rules for addition and subtraction of negative numbers.

When introducing complex numbers—which he regards as rather "sophistic"—Bombelli proceeds in much the same way as in the case of his definition of negative numbers: he introduces them formally by a "multiplication table". He calls $+\sqrt{-1}$, which can be "neither positive nor negative", piu di meno (plus from minus), and calls $-\sqrt{-1}$ meno di meno (minus from minus). Then he sets down the following table:

> piu di meno via piu di meno fa meno
> piu di meno via meno di meno fa piu
> meno di meno via piu di meno fa piu
> meno di meno via meno di meno fa meno
> (Ibidem, pp. 133–134).

Next Bombelli considers arithmetic operations on the new numbers. Thus he multiplies $\sqrt[3]{2+\sqrt{-3}} \, \sqrt[3]{2-\sqrt{3}}$ (which he writes as: Moltiplichisi, R.c. $\lfloor 2$ piu di meno R.q.3$\rfloor$ per R.c. $\lfloor 2$ meno di meno R.q.3$\rfloor$; here $\lfloor$ and $\rfloor$ play the role of our parentheses (and)), adds $a\sqrt{-1} \pm b\sqrt{-1}$, raises $a + b\sqrt{-1}$ to the second and third powers, and so on.

Bourbaki notes that Bombelli regards it as an axiom that piu and piu di meno cannot be added (in the sense of reduction of similar terms), and that this is one of the first occurrences of the notion of linear independence.

The second part of *Algebra* deals with the solution in terms of radicals of linear, quadratic, cubic, and quartic equations. Here Bombelli introduces his symbols $\underset{\smile}{1}, \underset{\smile}{2}, \underset{\smile}{3}, \ldots$ for powers of the unknown and, like Diophantus, introduces by means of tables rules for their multiplication: $\underset{\smile}{m} \, \underset{\smile}{n} = \underset{\smile}{m+n}$. Then he considers rules of operation on monomials $A^{\underline{n}}$, A, n concrete numbers, and polynomials. The exposition is clear and consistent. All he needs in order to obtain the supply of required formulas—such as $(a \pm b)^2$, $(a+b)(a-b)$, and so on—is literal notations for arbitrary parameters.

Bombelli's account of the solution of algebraic equations of the first four degrees is also consistent and systematic. When dealing with cubic equations he first explains the "irreducible" case. To this end he considers the cubic radicals in Cardano's formula. Until recently it has been thought (see, for example, [10]) that Bombelli found the value of radicals of the form $\sqrt[3]{a \pm i\sqrt{b}}$ accidentally. But this is not so. He proposed an original method for finding the value $\xi + i\sqrt{\eta}$ of $\sqrt[3]{a \pm i\sqrt{b}}$ or for finding bounds on ξ and η (see G. S. Smirnova's 1989 dissertation *From the History of Algebra in the 16th Century*).

Specifically, Bombelli puts $\sqrt[3]{a + i\sqrt{b}} = \xi + i\sqrt{\eta}$. Then $a + i\sqrt{b} = \xi^3 - 3\xi\eta + i(3\xi^2\sqrt{\eta} - \eta\sqrt{\eta})$. Hence $a = \xi^3 - 3\xi\eta$.

On the other hand, $\sqrt[3]{a - i\sqrt{b}} = \xi - i\sqrt{\eta}$. Hence $\xi^2 + \eta = \sqrt[3]{a^2 + b}$.

Since ξ and η are positive, Bombelli obtains for their determination two inequalities:

$$\xi^3 > a \quad \text{and} \quad \xi^2 < \sqrt[3]{a^2 + b}. \tag{4}$$

If ξ and η are integers, then their values can be determined in a finite number of trials. For example, in case of the equation $x^3 = 15x + 4$ Cardano's formula yields $\sqrt[3]{2 + i\sqrt{121}} + \sqrt[3]{2 - i\sqrt{121}}$. If we put $\sqrt[3]{2 + i\sqrt{121}} = \xi + i\sqrt{\eta}$, then ξ must satisfy the inequalities $\xi^3 > 2$ and $\xi^2 < \sqrt[3]{a^2 + b} = 5$. It follows that the only possible integral value of ξ is 2. Hence

$$x = (2 + i) + (2 - i) = 4.$$

If ξ and η are not integers, then the inequalities (4) provide bounds for them.

Of course Bombelli's method could not be used to find ξ and η in the general case because, in general, there is no algebraic procedure for obtaining a cube root of a complex number. Nevertheless, Bombelli managed to explain the mechanism for obtaining a solution in the "irreducible" case.

Finally, Part III of *Algebra* contains 272 problems with solutions of which, as mentioned earlier, 143 are taken from the first five books of Diophantus' *Arithmetica*. Diophantus' problems are interspersed with other problems that sometimes complement them and sometimes are unrelated to them. Bombelli provided detailed solutions of some of Diophantus' problems whose solutions were barely sketched by the latter. When Bombelli had a perfectly clear understanding of the method of solution of one of Diophantus' problems he changed Diophantus' data. He was the first to solve the following problem of Diophantus: To represent the difference of two cubes as a sum of two cubes:

$$x^3 + y^3 = a^3 - b^3, \quad a > b.$$

Diophantus claimed that the problem was always solvable and wrote down a [rational] solution for $a = 4$ and $b = 3$ without explaining how he obtained the values of x and y. Bombelli provided a solution for the very same values of the parameters but did not consider the general case.

In summary, we can say that Bombelli was the first European mathematician who appraised and creatively utilized Diophantus' algebraic methods and appreciated their obvious superiority over the tradition that went back to Leonardo Pisano. Following in the footsteps of Diophantus who introduced negative numbers, Bombelli introduced complex numbers and used them to solve algebraic equations.

3. François Viète

In a eulogy of Viète, de Thou, a well-known French historian and statesman, wrote: "François Viète, a native of Fontenay in Poitou, was a man of such immense genius and of such profundity of thought that he managed to reveal the innermost secrets of the most arcane sciences and easily managed to do all that human perspicacity is capable of. But of all the different studies that forever occupied his great and unwearied mind, the one he primarily applied his proficiency to was mathematics. So great was his mathematical distinction that all that the ancients had invented in this discipline, all that we missed as a result of the ravages of time that annihilated their creations, all these he reinvented, reintroduced, and enriched with much that was new. He thought so persistently that he would often spend three successive days in his study without food or sleep, except that from time to time he would rest his head on his arm for a brief spell of sleep to keep up his strength" (De Thou, Historiarum sui temporis continuatio, Frankfourti, 1625, Vol. 3, pp. 1003–1005).

We interrupt de Thou's colorful narrative and rely on dry prose for an account of the main events in Viète's life.

François Viète was born in Fontenay-le-Comte, some 60 kilometers from the famous Huguenot fortress La Rochelle. He was the son of an attorney. He studied law at the University of Poitiers and worked as a lawyer in his home town. But four years later he became secretary to the distinguished Huguenot courtier de Parthenay and tutor of his 12-year-old daughter Catherine. He studied cosmogony with his young student and was fascinated by the study of astronomy and trigonometry. It seems that already at that time he managed to express $\sin mx$ and $\cos mx$ as polynomials in $\sin x$ and $\cos x$. After the death of de Parthenay and the marriage of Catherine he followed his pupil to Paris. In 1571 he was appointed counselor to the *parlement* (court). Then he served as privy counselor to kings Henry III and Henry IV.

When Henry IV was at war with Spain, Viète rendered a very important service. To maintain the secrecy of their communications with the colonies and with the Netherlands the Spaniards used codes that contained more than 500 symbols. While the French managed to intercept their letters they were unable to decipher them. The king turned for help to Viète, who deciphered the letters without difficulty and continued to do so for two years. The Spaniards were stunned and publicly stated in Rome that the French king had used magic.

Viète lived in a period of bloody religious wars. His closeness to distinguished Huguenot families aroused the suspicion of fervid and powerful Catholics and at the end of 1584 he was banished from the court. He was recalled only at the beginning of 1589. These four years were a remarkably creative period for Viète. During this time he devoted his efforts to a major work, his *The Art of Analysis, or New Algebra*. In spite of his great diligence, so vividly described by de Thou, this work was not completed.

Before discussing Viète's fundamental works on algebra we mention a triumph that came to him very late in life. In 1593 the Netherlandish mathematician Adriaen van Roomen (1561–1615), or Romanus, published a treatise in which he computed π to 17 decimal places and challenged mathematicians to solve the equation

$$x^{45} - 45x^{43} + 945x^{41} - \cdots - 3795x^3 + 45x = A,$$

where

$$A = \sqrt{1\frac{3}{4} - \sqrt{\frac{5}{16}} - \sqrt{1\frac{7}{8} - \sqrt{\frac{45}{64}}}}.$$

He also included, without any explanations, three values of x corresponding to three values of A. Thus for $A = \sqrt{2 + \sqrt{2 + \sqrt{2 + \sqrt{2}}}}$,

$$x = \sqrt{2 + \sqrt{2 + \sqrt{2 + \sqrt{3}}}}.$$

The Netherlandish ambassador told Henry IV about the challenge and remarked that France had no mathematician capable of solving the problem. The king summoned Viète and informed him of the challenge. Viète immediately produced one solution and 22 more a day later (the remaining 22 solutions are negative). We will come back to Viète's method of solution in the sequel. At this point we add that he sent his solution to van Roomen together with a copy of his work *Apollonius Gallus, or The Restored Geometry of "Tangencies" by Apollonius of Perga*. This work was published in Paris in 1600. According to de Thou, the two works so impressed van Roomen that he traveled to France to meet Viète and to seek his friendship.

Viète died on 23 February 1603.

Viète had two prominent students. One was Marino Gethaldi (1566–1627) from Dubrovnik. The other was the Scotsman Alexander Anderson (1582–1619), who introduced a number of supplements into Viète's works and provided proofs of some of his theorems. Viète's collected works were published by Frans van Schooten in Leiden in 1646.

4. Creation of a literal calculus

Viète tried to create a new science (he called it ars analytica, or analytic art) that would combine the rigor of the geometry of the ancients with the operativeness of algebra. This analytic art was to be powerful enough to leave no problem unsolved: nullum non problema solvere.

Viète set down the foundations of this new science in his *An Introduction to the Art of Analysis* (In artem analyticem isagoge) of 1591.

In this treatise he created a literal calculus, i.e., he introduced the language of formulas into mathematics. Before him literal notations were restricted to the unknown and its powers. Such notations were first introduced by Diophantus and were somewhat improved by mathematicians of the 15th and 16th centuries (see Tables 1 and 2).

The first fundamentally new step after Diophantus was taken by Viète, who used literal notations for parameters as well as for the unknown. This enabled him to write equations and identities in general form. It is difficult to overestimate the importance of this step. Mathematical formulas are not just a compact language for recording theorems. After all, theorems can also be stated in words; for example, the formula

$$(a + b)^2 = a^2 + 2ab + b^2 \tag{1}$$

can be expressed by means of the phrase "the square of the sum of two quantities is equal to the square of the first quantity, plus the square of the second quantity, plus twice their product." After all, shorthand also has the virtue of brevity. What counts is that we can carry out operations on formulas in a purely mechanical manner and obtain in this way new formulas and relations. To do this we must observe three rules: 1) the rule of substitution; 2) the rule for removing parentheses; and 3) the rule for reduction of similar terms. For example, from formula (1) one can obtain in a purely mechanical manner, without reasoning, formulas for $(a + b + c)^2$, for $(a + b)^3$, and so on. In other words, literal calculus replaces some reasoning by mechanical computations. In Leibniz' words, literal calculus "relieves the imagination".

Table 2
RECORDS OF EQUATIONS

1. Diophantus of Alexandria (*3rd century* CE)

$x^3 = 2 - x$ $\mathrm{K}^v \bar{\alpha} \; \acute{\iota}' \sigma \; \overset{\circ}{\mathrm{M}} \bar{\beta} \; \pitchfork \; \varsigma \bar{\alpha}$

$8x^3 - 16x^2 = x^3$ $\mathrm{K}^v \bar{\eta} \; \pitchfork \; \Delta^v \bar{\iota \varsigma} \; \acute{\iota}' \sigma \mathrm{K}^v \bar{\alpha}$

2. Luca Pacioli (*ca. 1445–ca. 1559*)

$x^2 + x = 12$ 1.ce.p̂.1.co.e q̂ le a 12.

3. Nicolas Chuquet (*d. 1500*)

$\sqrt{3x^4 - 24} = 8$ R_x^2. 3^4.m̂.24 est egale a 8

4. Michael Stifel (*1486–1567*)

$\dfrac{116 + \sqrt{41472}}{-18x - \sqrt{648}x} = 0$ $116 + \sqrt{_3}41472 - 18\mathfrak{r} - \sqrt{_3}648\mathfrak{r}$ aequantur 0

5. Girolamo Cardano (*1501–1576*)

$x^3 = 15x + 4$ 1.*cu.aequalis*15.*rebus* p̂.4.

6. Rafael Bombelli (*ca. 1526–1573*)

$x^6 - 10x^3 + 16 = 0$ 1.6̤ m.10 3̤ p̂.16 eguale a 0

7. François Viète (*1540–1603*)

$x^3 - 8x^2 + 16x = 40$ $1C - 8Q + 16N$ aequ. 40

$x^3 + 3bx = 2c$ *Acubus + Bplano3inA aequari Zsolido2*

8. Thomas Harriot (*1560–1621*)

$a^3 - 3ab^2 = 2c^3$ *aaa − 3bba = 2ccc*

9. Albert Girard (*1595–1632*)

$x^3 = 13x + 12$ 1③ × 13① + 12

10. René Descartes (*1596–1650*)

$x^3 + px + q = 0$ $x^3 + px + q^{\infty} 0$

We can hardly imagine mathematics without formulas, without a calculus. But it was such up until Viète's time. The importance of the step taken by Viète is so fundamental that we consider his reasoning in detail.

Viète adopted the basic principle of Greek geometry according to which only homogeneous magnitudes can be added, subtracted, and can be in a ratio to one another. As he put it: "Homogena homogenei comparare." As a result of this principle he divides magnitudes into "species": the 1st species consists of "lengths", i.e., of one-dimensional magnitudes. The product of two magnitudes of the 1st species belongs to the 2nd species, which consists of "plane magnitudes" or "squares", and so on.

In modern terms the domain V of magnitudes considered by Viète can be described as follows:

$$V = \mathbf{R}_+^{(1)} \cup \mathbf{R}_+^{(2)} \cup \ldots \cup \mathbf{R}_+^{(k)} \cup \ldots,$$

where $\mathbf{R}_+^{(k)}$ is the domain of k-dimensional magnitudes, $k \in \mathbf{N}_+$. In each of the domains $\mathbf{R}_+^{(k)}$ we can carry out the operations of addition and of subtraction of a smaller magnitude from a larger one, and can form ratios of magnitudes. If $\alpha \in \mathbf{R}_+^{(k)}$ and $\beta \in \mathbf{R}_+^{(l)}$, then there is a magnitude $\gamma = \alpha\beta$ and $\gamma \in \mathbf{R}_+^{(k+l)}$. If $k > l$, then there exists a magnitude $\delta = \alpha : \beta$, and $\delta \in \mathbf{R}_+^{(k-l)}$.

After constructing this "ladder" of magnitudes Viète proposes to denote unknown magnitudes by vowels $A, E, I, O \ldots$ and known ones by consonants $B, C, D, \ldots$. Furthermore, to the right of the letter denoting a magnitude he places a symbol denoting its species. Thus if B is in $\mathbf{R}_+^{(2)}$, then he writes B plan. (i.e., planum—plane), and if an unknown A is in $\mathbf{R}_+^{(2)}$, then he writes A quad. (square). Similarly, magnitudes in $\mathbf{R}_+^{(3)}$ are indexed solid or cub, and those in $\mathbf{R}_+^{(4)}$ are indexed plano-planum or quadrato-quadratum, and so on.

For addition and subtraction Viète adopts the cossist symbols $+$ and $-$, and introduces the symbol $=$ for the absolute value of the difference of two numbers; thus $B = D$ is the same as $|B - D|$. For multiplication he uses the word "in", A in B, and for division the word "applicare".

Next he introduces the rules

$$B - (C \pm D) = B - C \mp D; \quad B \text{ in } (C \pm D) = B \text{ in } C \pm B \text{ in } D,$$

as well as operations on fractions written by means of letters, e.g.,

$$\frac{Bpl}{D} + Z = \frac{Bpl + ZinD}{D}.$$

Viète's next treatise was *Ad logisticam speciosam notae priores*, which appeared only in 1646 as part of his collected works. In it he set down some of the most important algebraic formulas, such as:

$$(A + B)^n = A^n \pm nA^{n-1}B + \cdots \pm B^n, \quad n = 2, 3, 4, 5;$$

$$A^n + B^n = (A + B)(A^{n-1} - A^{n-2}B + \cdots \pm B^{n-1}), \quad n = 3, 5;$$

$$A^n - B^n = (A - B)(A^{n-1} + A^{n-2}B + \cdots + B^{n-1}), \quad n = 2, 3, 4, 5.$$

Viète's literal calculus was perfected by René Descartes (1596–1650). He dispensed with the principle of homogeneity and gave the literal calculus its modern form. At the end of the 17th century a calculus was created for the analysis of infinitesimals (it was called for a long time the "algebra of the infinite"). One of its forms was Newton's method of fluxions and infinite series (a generalization of polynomials!) and the other was the differential and integral calculus of Leibniz. Variational calculi and a calculus of partial differentials and derivatives were developed in the 18th century. A calculus of logic was created in the 19th century. Today, almost every mathematical theory has its own calculus (vector calculus, tensor calculus, and so on); furthermore, special calculi are created for individual problems, both theoretical and practical. The apparatus of formulas has become an indispensable language of mathematics. And its originators were Diophantus and Viète.

5. Genesis triangulorum

Genesis triangulorum is the title of the last part of *Ad logisticam speciosam notae priores*. It contains 12 propositions (XLV–LVI). In the first nine of these propositions Viète constructs a calculus of triangles based on the formula for composition of forms

$$(x^2+y^2)(u^2+v^2) = (xu-yv)^2+(xv+yu)^2 = (xu+yv)^2+(xv-yu)^2 \quad (5)$$

(see Ch. 4). Viète associates with the form $x^2 + y^2$ a right triangle with base x, height y, and hypotenuse $\sqrt{x^2 + y^2}$ (we will denote it by (x, y, z)) and interprets (5) as the formula for the "composition" of triangles (x, y, z) and (u, v, w). Then he poses the problem: "To construct (effingere) from two right triangles a third right triangle."

He explains that the hypotenuse of the third triangle must be equal to the product of the hypotenuses of the given triangles.

According to formula (5), there are two ways of constructing the required third triangle from the given triangles (x, y, z) and (u, v, w):

$$(x, y, z) \otimes (u, v, w) = \begin{cases} 1) \, (|xu - yv|, xv + yu, zw), \\ 2) \, (xu + yv, |xv - yu|, zw) \end{cases}$$

(Figure 18).

Viète calls the triangle obtained from the first of these operations synaereseos (from the Greek $\sigma\upsilon\nu\alpha\iota\rho\acute{\epsilon}\omega$, to combine) and the triangle obtained from the second of these operations diareseos (from the Greek $\delta\iota\alpha\iota\rho\acute{\epsilon}\omega$, to section). He does not explain the reason for these names but writes that he will give it "at the appropriate place." We can infer from his other works and from remarks of his student Anderson that Viète named the resulting triangles the way he did because he realized that the acute base angle of the first triangle

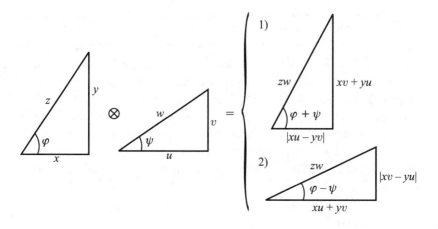

FIGURE 18

was equal to the sum of the acute base angles of the composed triangles, and that of the second triangle, to their difference.

Thus the operation of composition of triangles splits into two, to be denoted by $\otimes_1$ and $\otimes_2$ respectively. We will try to explain their significance from the viewpoint of modern mathematics.

We establish the connection between the operations of composition and multiplication of complex numbers. To this end we associate with the right triangle (x, y, z) the complex number $\alpha = x + iy$ with norm $N(\alpha) = x^2 + y^2$ and argument $\varphi = \arctan \frac{y}{x}$. Conversely, with every complex number $\alpha = x + iy$, $x > 0$, $y > 0$, we can associate the Viète right triangle $\left(x, y, \sqrt{N(\alpha)} \right)$.

If we apply the operation $\otimes_1$ to two right triangles (x, y, z) and (u, v, w) with respective acute base angles φ and ψ such that $\varphi + \psi < \pi/2$, then, indeed, to this operation there corresponds the operation of multiplication of the complex numbers $\alpha = x + iy$ and $\beta = u + iv$.

If $\pi/2 < \varphi + \psi < \pi$, then Viète obtains by composition of forms $xu - yv < 0$, i.e., from our viewpoint, a complex number in the second quadrant. But he associates with it the triangle with sides $(|xu - yv|, xv + yu, zw)$, i.e., he goes over from the number $-a + bi$ ($a > 0, b > 0$) to the number $a + bi$. It is easy to see that the result of the first operation can be written as follows:

$$\alpha \otimes_1 \beta = \begin{cases} \alpha\beta, & \text{if } 0 < \varphi + \psi < \pi/2, \\ -\overline{\alpha\beta}, & \text{if } \pi/2 < \varphi + \psi < \pi. \end{cases}$$

Viète makes very clever use of the operation $\otimes_1$ to multiply more than two factors. Specifically, to apply $\otimes_1$ to triangles (x, y, z), (u, v, w), and

(p, q, r) he composes successively the forms $x^2 + y^2$, $u^2 + v^2$, and $p^2 + q^2$.
Let the resulting form be $a^2 + b^2$. These compositions correspond to the
multiplication $(x+iy)(u+iv)(p+iq) = a+bi$; $N(a+bi) = a^2 + b^2$, where,
of course, a and b can be positive or negative. Viète interprets the final result
as the triangle $\left(|a|, |b|, \sqrt{a^2 + b^2}\right)$. This triangle represents the four complex
numbers $a + bi$, $-a + bi$, $a - bi$, and $-a - bi$. This awkward aspect of
Viète's calculus is due to his unwillingness to use negative numbers. No
misunderstandings arise, however, because the triangle $\left(|a|, |b|, \sqrt{a^2 + b^2}\right)$—
call it "reduced"—usually turns up only in connection with the interpretation
of final results.

Similarly, the second operation of composition of triangles can be defined
by the relations

$$\alpha \otimes_2 \beta = \begin{cases} \alpha\overline{\beta}, & \text{if } \varphi - \psi > 0, \\ \overline{\alpha\beta}, & \text{if } \varphi - \psi < 0, \end{cases}$$

i.e., this operation corresponds to the multiplication of a complex number by
$\overline{\beta} = u - iv$ and the subsequent interpretation in terms of a "reduced triangle".
This operation is neither associative nor commutative.

We note that every right triangle (x, y, z) is uniquely determined by its
hypotenuse z and its acute base angle φ. Thus one can associate with a Viète
triangle not only the algebraic form $x + iy$ of a complex number but also its
polar form $z(\cos\varphi + i\sin\varphi)$.

In propositions XLVIII–LI the first composition operation is applied to
two equal triangles (a, b, c), then to the resulting triangle (a_1, b_1, c_1) and the
initial triangle (a, b, c), and so on. We could say that he is raising the triangle
(a, b, c) to a positive integral power n, which is equivalent to raising $a + bi$
or $c(\cos\varphi + i\sin\varphi)$ to the power in question. We consider these propositions
in greater detail.

In proposition XLVIII Viète finds $(a, b, c) \otimes_1 (a, b, c) = (a^2 - b^2, 2ab, c^2)$.
He notes that the acute base angle of the resulting triangle is $\varphi_1 = 2\varphi$
and therefore calls it a double-angle triangle. (We interpolate a remark. If
$2\varphi < \pi/2$ then $a^2 - b^2 > 0$, and Viète's assertion needs no supplementary
arguments. But if $\pi/2 < 2\varphi < \pi$, then $a^2 - b^2 < 0$. If we take the reduced
triangle $(|a^2 - b^2|, 2ab, c^2)$, then its acute base angle is $\pi - 2\varphi$. This kind of
reduction will have to be done in the sequel.) In proposition XLIX he applies
the composition $\otimes_1$ to the initial triangle and to the double-angle triangle and
obtains the triangle $(|(a^3 - 3ab^2|, |3a^2b - b^3|, c^3)$, which he calls a triple-angle
triangle. It is not difficult to see that this third triangle is the result of the

composition of forms:

$$[(a^2 - b^2)^2 + (2ab)^2](a^2 + b^2).$$

Similarly, in proposition L he composes the initial triangle and the triple-angle triangle and obtains the triangle

$$(|a^4 - 6a^2b^2 + b^4|, \ |4a^3b - 4ab^3|, \ c^4).$$

Finally, in proposition LI, he obtains in the same way a triangle with base angle 5φ and sides

$$(|a^5 - 10a^3b^2 + 5ab^4|, \ |5a^4b + 10a^2b^3 - b^5|, \ c^5).$$

Now Viète formulates a general rule of "separation" (diductio) of right triangles:

"From this one obtains a general consequence on separation of right triangles.

If one forms an arbitrary power of a binomial and the resulting individual homogeneous terms are successively separated into two parts and in both are taken first positive and then negative, then the base of a certain right triangle will be similar to the first part and its height to the second. And the hypotenuse will be similar to the power itself.

That same triangle, whose base is similar or equal to one of the roots of the composed (binomial) and height to another, is named after its angle, subtended by the height. Indeed, it is convenient to name the triangles, obtained by separation of these very roots by raising to an arbitrary power, according to the multiplicity of that very angle. Namely: double if the power is a square, triple if the power is a cube, quadruple if the power is a square–square, quintuple if the power is a square–cube, and in that same sequence to infinity."

Thus Viète's rule is equivalent to the formula for raising $a + bi$ to an arbitrary positive integral power $(a + bi)^n$.

Indeed, Viète is telling us to raise the binomial $a + b$ to the nth power,

$$(a+b)^n = a^n + na^{n-1}b + \frac{n(n-1)}{1 \cdot 2}a^{n-2}b^2 + \frac{n(n-1)(n-2)}{1 \cdot 2 \cdot 3}a^{n-3}b^3 + \cdots,$$

and to separate the resulting homogeneous terms into two series whose terms are given alternating signs:

$$1) \quad a^n - \frac{n(n-1)}{1 \cdot 2}a^{n-2}b^2 + \cdots,$$

$$2) \quad na^{n-1}b - \frac{n(n-1)(n-2)}{1 \cdot 2 \cdot 3}a^{n-3}b^3 + \cdots$$

Then the first series will correspond to the base of the resulting triangle and the second to its height. In other words,

$$\mathrm{Re}(a + bi)^n = a^n - \frac{n(n-1)}{1 \cdot 2} a^{n-2} b^2 + \cdots,$$

$$\mathrm{Im}(a + bi)^n = na^{n-1}b - \frac{n(n-1)(n-2)}{1 \cdot 2 \cdot 3} a^{n-3} b^3 + \cdots$$

Viète does not say that one must take $|\mathrm{Re}(a + bi)^n|$ and $|\mathrm{Im}(a + bi)^n|$ for the resulting reduced triangle. Clearly, his method makes this necessary.

In the second half of his conclusion Viète notes that the hypotenuse of the resulting triangle is z^n and the base angle is $n\varphi$. In other words,

$$[z(\cos\varphi + i\sin\varphi)]^n = z^n(\cos n\varphi + i\sin n\varphi).$$

Thus this theorem includes the so-called de Moivre formula. We note that knowing this formula Viète could immediately solve van Roomen's problem (see 3). Indeed, using the de Moivre formula he could express $\sin n\varphi$ and $\cos n\varphi$ as polynomials in $\sin\varphi$ and $\cos\varphi$.

Viète noticed that the given value of A is an expression for the side of a regular 15-gon inscribed in a unit circle (i.e., for the chord of an arc of $24°$). The coefficients of the equation showed that it expresses $\sin\varphi$ in terms of $\sin(\varphi/45)$, i.e., x must be the chord of $1/45$ of this arc or, equivalently, it must subtend an arc of $(8/15)°$. But then $x = 2\sin(4/15)°$. Since what was required was a geometric solution, the problem was solved. But Viète gave 22 more solutions:

$$x_k = 2\sin\left(\frac{12° + 360°k}{45}\right), \qquad k = 1, \ldots, 22$$

(the remaining 22 solutions are negative, so Viète ignored them).

It is safe to say that Viète constructed an impeccably rigorous and original calculus of triangles equivalent to the multiplication of complex numbers, and that he derived a formula for raising a complex number—in its usual form $(a + bi)$ or in its polar form $(r(\cos\varphi + i\sin\varphi))$—to an arbitrary positive integral power. He did this without introducing new "objects" or "symbols" such as $\sqrt{-1}$.

We compare Bombelli's complex-number symbols with Viète's calculus of triangles. Each of the two systems has its advantages and disadvantages. Bombelli's complex-number symbols were convenient for carrying out the four arithmetical operations; in modern terms, they formed a field, i.e., they "behaved as well" under the two composition laws defined for them as did the rational numbers under addition and multiplication. However, there was no analog of the polar form of a complex number for these symbols, i.e., no notions of modulus and of argument. Hence they were not convenient when it came to extraction of roots or to trigonometric applications.

Viète's calculus of triangles admitted an algebraic as well as a trigonometric interpretation and it was therefore immediately utilized to obtain key trigonometric formulas. Viète also used it for solving indeterminate equations. But this calculus was not very operative. Also, the only operation defined for triangles was composition, the analog of multiplication, whereas, as already noted, Bombelli's number symbols could be added and multiplied and formed a field under these operations. That is why, during the further development of mathematics in the modern era, the more operative number-symbols of Bombelli won the day.

But there were mathematicians in the 17th century who used Viète's triangle-numbers and preferred them to Bombelli's number-symbols introduced without any justification. Thus Fermat used exclusively Viète triangles, e.g., in connection with the problem of finding a number that would be the hypotenuse of a right triangle a given number of times (see Fermat's letter to Frenicle dated 15 June 1641 (P. Fermat, *Oeuvres,* ed. P. Tannery et Charles Henry, Paris, Gauthier-Villars, 1891–1912, letter No. XLVIII)).

In the 18th century mathematicians began to write Bombelli's number-symbols in polar form, and in the 19th century these number-symbols entered analysis. Their adoption was promoted by Gauss' construction of the arithmetic of complex numbers—a development that made them into "genuine numbers".

6. Indeterminate equations in the work of Viète

In his work *Zetetica*, Viète used literal calculus and the calculus of triangles to solve indeterminate equations, and in subsequent works—to solve determinate equations.

The first five problems in Book IV of *Zetetica* (with the exception of theorem 4) reduce to the equations

$$x^2 + y^2 = B^2; \tag{6}$$

$$x^2 + y^2 = B^2 + D^2. \tag{7}$$

We saw that these equations were first solved by Diophantus (see Ch. 3) and then—in a different way—by Leonardo Pisano (see Ch. 4). Essentially, Viète reproduces both of these solutions. What is new in his treatment of equations (6) and (7) is that he was the first to write them as equalities involving literal expressions, and this enabled him to solve them "in general form". We give his solution of problem (7).

Without mentioning Leonardo he first gives a solution based on the latter's method. He takes a right triangle (B, D, Z), where $Z = \sqrt{B^2 + D^2}$ can be irrational, and another right triangle (p, q, r) with rational sides. Then

$$(B, D, Z) \otimes_1 (p, q, r) = (|Bp - Dq|, Bq + Dp, rZ);$$

$$(B, D, Z) \otimes_2 (p, q, r) = (Bp + Dq, |Bq - Dp|, rZ).$$

Both operations are carried out according to the rules stated in *Genesis triangulorum* and yield two right triangles with the same hypotenuses rZ. Then Viète takes triangles similar to these two:

$$1) \quad \left(\frac{|Bp - Dq|}{r}, \quad \frac{Bq + Dp}{r}, \quad Z \right);$$

$$2) \quad \left(\frac{Bq + Dp}{r}, \quad \frac{|Bq - Dp|}{r}, \quad Z \right),$$

and these yield directly two new solutions of equation (7):

$$x_1 = \frac{|Bp - Dq|}{r}; \quad y_1 = \frac{Bq + Dp}{r}; \tag{8}$$

$$x_2 = \frac{Bq + Dp}{r}; \quad y_2 = \frac{|Bq - Dp|}{r}. \tag{9}$$

In terms of complex numbers, the first of these solutions is obtained by multiplying $\alpha = B + Di$ by $\beta = \frac{p}{r} + i\frac{q}{r}$, where $\frac{p}{r}$ and $\frac{q}{r}$ are rational and $N(\beta) = 1$, and the second, by multiplying $\overline{\alpha} = B - Di$ by β.

Viète takes the second solution method from Diophantus but first algebraicizes it. He puts the first of the required unknowns equal to $A + B$, the second to $(S/R)A - D$, substitutes these expressions in (7), and obtains

$$x = A + B = \frac{2RSD + B(S^2 - R^2)}{S^2 + R^2};$$

$$y = \frac{|(S^2 - R^2)D - 2RSB|}{S^2 + R^2}. \tag{10}$$

Then he notes that the expressions (10) and (9) coincide if one supposes that $p = S^2 - R^2$, $q = 2RS$, and $r = S^2 + R^2$.

In problems 18–20 of Book IV of *Zetetica*, Viète turns to the problem of four cubes. He successively solves the indeterminate equations:

$$x^3 + y^3 = B^3 - D^3, \quad B > D; \tag{IV$_{18}$}$$

$$x^3 - y^3 = B^3 + D^3; \tag{IV$_{19}$}$$

$$x^3 - y^3 = B^3 - D^3, \quad B > D. \tag{IV$_{20}$}$$

Problem IV$_{18}$ was formulated by Diophantus in connection with his solution of problem V$_{16}$ in his *Arithmetica*. He claimed that it is always solvable. We saw (in §2, Ch. 5) that Bombelli solved it for $B = 4, D = 3$. But he too was unable to prove its solvability for arbitrary $B > D$. This was first done by Fermat. Readers interested in the "Problem of four cubes" are referred to *The History of Diophantine Analysis from Diophantus to Fermat* [2] (Russian).

7. Beginning of the theory of determinate equations

In his treatise *On Perfecting Equations* Viète embarked on a systematic investigation of equations with literal coefficients. His first step was to transform equations by means of substitutions. He noted that given the equation

$$x^n + a_1 x^{n-1} + \cdots + a_{n-1}x + a_n = 0 \tag{11}$$

(we are using modern symbolism), it is possible to remove its second term by means of the substitution

$$x = y - a_1/n. \tag{12}$$

Next, using the substitution $x = a/y$, it is possible to permute the terms of the equation in a manner Viète called "last—first". Finally, using the substitution $x = ky$, it is possible to obtain an equation all of whose rational roots are integers.

Viète formulated the theorem on the connection between the coefficients of an equation and its roots. He limited himself to equations of degrees $n = 2, \ldots, 5$. This theorem is named after him. It asserts that

$$x_1 + x_2 + \cdots + x_n = -a_1$$
$$x_1x_2 + x_1x_3 + \cdots + x_{n-1}x_n = a_2,$$
$$\cdots\cdots\cdots\cdots\cdots\cdots\cdots\cdots\cdots\cdots \tag{13}$$
$$x_1x_2 \cdots x_n = (-1)^n a_n.$$

In his book *New Discoveries in Algebra*, published in 1629, Albert Girard formulated Viète's theorem for an equation of arbitrary degree.

Viète's ideas launched two very important types of investigations in the theory of equations. One was related to the substitution (12) and could be formulated as the question: Is it possible to eliminate the second and third terms by means of a substitution $x = y^2 + Ay + B$? More generally, is it possible to choose the coefficients of the substitution $x = y^{n-1} + A_1 y^{n-2} + \cdots + A_{n-1}$ so as to eliminate all intermediate terms, i.e., so as to reduce the equation to the form

$$y^n \pm C = 0,$$

and thus express its roots by radicals? The first to follow this approach was Leibniz' friend Tschirnhaus, whose work we will discuss in the sequel.

The other kind of investigation involved the study of symmetric functions of the roots of an equation. The functions of the roots in (13) that give the coefficients of the equation are called elementary symmetric functions. Already Girard, and then Newton, considered symmetric functions of the form $S_m = \sum_{i=1}^n x_i^m$. For $m = 1, 2, 3, 4$ Girard expressed them in his book in terms of the coefficients of the original equation:

$$S_1 = -a_1, \quad S_2 = a_1^2 - 2a_2, \quad S_3 = -a_1^3 + 3a_1 a_2 - 3a_3,$$

$$S_4 = a_1^4 - 4a_1^2 a_2 + 4a_1 a_3 + 2a_2^2 - 4a_4,$$

and Newton, in his *Arithmetica universalis* of 1707, gave recursion formulas for finding sums of powers of the roots:

$$S_m + a_1 S_{m-1} + a_2 S_{m-2} + \cdots + m a_m = 0.$$

By means of these formulas it is possible to express the sums S_m as polynomials with integral coefficients in terms of the coefficients of the equation (and thus in terms of the elementary symmetric functions). This (Girard-Newton) theorem about sums of powers of the roots of an equation was the first step leading to one of the fundamental results of the theory of equations which asserts that every rational symmetric function of the roots of an equation is a rational function of the elementary symmetric functions and thus of the coefficients of the equation.

We conclude this section with a discussion of Viète's work on cubic equations.

Viète gave the following elegant solution of the cubic equation

$$x^3 + 3ax = 2b \tag{14}$$

(of course, he used different symbols for the unknowns and the parameters). He put

$$a = t^2 + xt = t(t + x), \tag{15}$$

probably with a view to reducing equation (14) to the form

$$(x+t)^3 - t^3 = 2b.$$

However, elimination of x from (14) and (15) yields immediately a quadratic equation in t^3:

$$(t^3)^2 + 2bt^3 = a^3.$$

In his *Supplement to Geometry* Viète showed that the "irreducible case" of a cubic equation reduces to angle trisection and thus admits of a trigonometric solution.

Indeed, consider the equation

$$x^3 - px = q, \quad \text{with} \quad \left(\frac{q}{2}\right)^2 < \left(\frac{p}{3}\right)^3. \tag{16}$$

We rewrite it as

$$x^3 - 3r^2 x = ar^2. \tag{17}$$

Since $(ar^2/2)^2 < (r^2)^3$, i.e., $a < 2r$, we can put $a = 2r\sin v$ and write equation (17) as

$$(x/r)^3 = 3(x/r) + 2\sin v. \tag{17'}$$

Putting $x/r = -y$, Viète obtains the equation

$$3y - y^3 = 2\sin v. \tag{18}$$

In view of the well-known relation

$$3\sin\varphi - 4\sin^3\varphi = \sin 3\varphi,$$

he obtains the equation

$$3(2\sin\varphi) - (2\sin\varphi)^3 = 2\sin 3\varphi. \tag{19}$$

From this he obtains $y_1 = 2\sin(v/3)$. The second positive root is $y_2 = 2\sin\left(\frac{v+2\pi}{3}\right)$. The third root is negative. Viète obtained all expressions for the roots geometrically.

Essentially, this argument showed that in the "irreducible" case a cubic equation has three different real roots. Since Viète admitted only positive roots he could not formulate this conclusion explicitly. This was done by Girard in his *New Discoveries in Algebra*.

In summary, we can say that the end of the 16th century marked a crucial turning point in the evolution of algebra, for at that time it found its own language, namely the literal calculus. This made it possible to conduct general investigations of determinate and indeterminate equations. At that time the domain of numbers was extended to the field **C** of complex numbers. While not yet entirely "legitimate", these numbers were used to advantage in algebraic investigations. This was the basis for the subsequent development of algebra, which flourished in the second half of the 18th century and in the 19th century.

CHAPTER 6

Algebra in the 17th and 18th centuries

1. The arithmetization of algebra

The 16th century was marked by remarkable achievements in algebra and was followed by a period of relative calm in this area. The attention of 17th-century mathematicians was primarily directed towards infinitesimal analysis, which was created at that time. Nevertheless, while inconspicuous at first sight, profound changes were taking place in algebra that can be characterized by one word—arithmetization.

The first steps in this direction were taken by the famous philosopher and mathematician René Descartes (1596–1650). In his *Geometry* (the fourth part of his 1637 *Discourse on Method*), whose essential content was the reduction of geometry to algebra, or, in other words, the creation of analytic geometry, he first of all transformed Viète's calculus of magnitudes (logistica speciosa). Descartes represented all magnitudes by segments and constructed a calculus of segments that differed essentially from the one that was used in antiquity and that formed the basis of Viète's construction. Descartes' idea was that the operations on segments should be a faithful replica of the operations on rational numbers. Whereas the ancients and Viète regarded the product of two segment magnitudes as an area, i.e., as a magnitude of dimension 2, Descartes stipulated that it was to be a segment. To this end, he introduced a unit segment—which we will denote by e—and defined the product of segments a and b as the segment c that was the fourth proportional to the segments e, a, and b. Specifically (see Figure 19), he constructed an arbitrary angle ABC and laid off the segments $AB = e$, $BD = b$, and $BC = a$. Then he joined A to C, drew $DF \parallel AC$, and obtained the segment $BF = c = ab$. This meant that the product belonged to the same domain of magnitudes (segments) as the factors. Division was defined analogously: to divide $BF = c$ by $BD = b$ we lay off

91

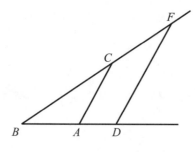

FIGURE 19

from the vertex B of the angle the segment $BA = e$, join F to D, and draw AC parallel to DF. The segment BC is the required quotient. In this way Descartes made the domain of segments into a replica of the semifield $\mathbf{R}^+$. Later he also introduced negative segments (with directions opposite to those of the positive segments) but did not go into the details of operations with negative numbers. Finally, Descartes showed that the operation of extraction of square roots (of positive magnitudes) does not take us outside the domain of segments. (We interpolate a comment. Already Bombelli introduced similar rules of operation with segments. Until recently it was thought that he did this in the fourth part of his manuscript published only in the 20th century. However, G. S. Smirnova showed recently (see the comment on p. 73) that such operations with segments occur also in the parts of Bombelli's *Algebra* published in 1572, i.e., during his lifetime.) To extract the square root of $c = BF$, Descartes extended this segment, laid off $FA = e$ on the extension, drew a semicircle with diameter BA, and erected at F the perpendicular to BA. If I is its intersection with the semicircle, then $FI = \sqrt{c}$.

Descartes' calculus was of tremendous significance for the subsequent development of algebra. It not only brought segments closer to numbers but also lent algebra the simplicity and operativeness which we take advantage of to this day. Another convention introduced by Descartes and used to this day is the denoting of unknowns by the last letters of the alphabet: x, y, z, and of knowns by the first letters: a, b, c. The only difference between Descartes' symbolism and modern symbolism is his equality sign: $\sim$o.

Essentially it was Descartes who established the isomorphism between the domain of segments and the semifield $\mathbf{R}^+$ of real numbers. However, he gave no general definition of number. This was done by Newton in his *Universal Arithmetic* in which the construction of algebra on the basis of arithmetic reached its completion. He wrote: "Computation is conducted either by means of numbers, as in ordinary arithmetic, or through general variables, as is the habit of analytical mathematicians." And further: "Yet arithmetic is so instrumental to algebra in all its operations that they seem jointly to constitute but a single, complete computing science, and for that reason I shall explain both together."

Newton immediately gives a general definition of number. We recall that in antiquity number denoted a collection of units (i.e., natural numbers), and that ratios of numbers (rational numbers) and ratios of like quantities (real numbers) were not regarded as numbers. Claudius Ptolemy (2nd century CE) and Arab mathematicians did identify ratios with numbers, but in 16th- and 17th-century Europe the Euclidean tradition was still very strong. Newton was the first to break with it openly. He wrote: "By a 'number' we understand not so much a multitude of units as the abstract ratio of any quantity to another quantity which is considered to be unity. It is threefold: integral, fractional, and surd. An integer is measured by unity, a fraction by a submultiple part of unity, while a surd is incommensurable with unity."

With characteristic brevity, Newton goes on to define negative numbers: "Quantities are either positive, that is, greater than zero, or negative, that is, less than zero. . . .in geometry, if a line drawn with advancing motion in some direction be considered as positive, then its negative is one drawn retreating in the opposite direction.

To denote a negative quantity . . . the sign $-$ is usually prefixed, to a positive one the sign $+$."

Then Newton formulates rules of operation with relative numbers. We quote his multiplication rule: "A product is positive if both factors are positive or both negative and it is negative otherwise."

He provides no "justifications" for these rules.

Thus Viète's elaborate domain of magnitudes was replaced in the 17th century by the field of real numbers and arithmetic formed the foundation of algebra.

2. Descartes' treatment of determinate equations

In the last part of his *Geometry* Descartes presents his treatment of equations. He consistently writes them with zero on the right "for this is often the best form to consider". This enables him to find the following properties of equations:

1) if α is a root of an equation then its left side is algebraically divisible by $x - \alpha$;

2) an equation can have as many positive roots as it contains changes of sign from $+$ to $-$; and as many false (i.e., negative) roots as the number of times two $+$ signs or two $-$ signs are found in succession;

3) in every equation one can eliminate the second term by a substitution;

4) the number of roots of an equation can be equal to its degree.

For cubic and quartic equations Descartes formulated without proof a number of nontrivial assertions.

Consider a cubic equation

$$x^3 + ax + b = 0, \quad a, b \in \mathbf{Q}. \tag{1}$$

Assuming that all its roots are real, Descartes investigates their constructibility by ruler and compass. He claims that for this it is necessary and sufficient that equation (1) be reducible over $\mathbf{Q}$, i.e., that it have a rational root α. Descartes proved the sufficiency of this condition as follows: upon division of the left side of the equation by $x - \alpha$ we obtain a quadratic equation whose roots are constructible by ruler and compass. The necessity of this condition is much harder to prove and Descartes says nothing about this matter.

Descartes considers the same question for a quartic. He claims that its roots are constructible by ruler and compass if and only if the auxiliary cubic (called the cubic resolvent; for more on this term see 5 below) used in Ferrari's solution of the quartic is reducible. Finally, Descartes discovered the method of undetermined coefficients, a method which has played a major role in algebra as well as in analysis (especially in the theory of series). Using this method he found a new way of solving a quartic equation (see the next section).

3. The fundamental theorem of algebra

This theorem, which became one of the central problems of 18th-century algebra, was formulated in the 17th century by Descartes and Girard. Since he did not take into account complex roots, Descartes formulated it cautiously: "Every equation can have as many distinct roots (values of the unknown quantity) as the number of dimensions of the unknown quantity in the equation." Girard overcame this psychological difficulty and stated in his *New Discoveries in Algebra* of 1629 that the number of solutions of an algebraic equation is equal to its degree. Girard operated freely with negative and complex roots.

In addition to Girard's formulation, 18th-century mathematicians used the following, equivalent, version of the fundamental theorem: every polynomial

$$f_n(x) = x^n + a_1 x^{n-1} + \cdots + a_{n-1} x + a_n \tag{2}$$

with real coefficients can be written as a product of linear and quadratic factors with real coefficients.

The first proof of this theorem was given in 1746 by d'Alembert (1717–1783). It was purely analytic and its standards of rigor were low even for the 18th century. In the same year Euler (1707–1783), the greatest mathematician of the 18th century and a brilliant exponent of its ideas, presented his proof of the theorem to the Berlin Academy. It was subsequently included in the memoir *Investigations of Complex Roots of Equations* (Recherches sur les racines imaginaires des équations, 1751).

Unlike d'Alembert, Euler looked for an algebraic proof of the fundamental theorem. Today we know that the theorem cannot be proved without the use of certain continuity assumptions. It appears that Euler was aware of this. In his proof he reduced the nonalgebraic assumptions to a minimum. He used the following two:

I. Every equation of odd degree with real coefficients has at least one real root.

II. Every equation of even degree with real coefficients and negative constant term has at least two real roots.

Euler's idea for the rest of his proof was to use a process that would reduce the solution of an equation of degree $2^k m$, m odd, to an equation of degree $2^{k-1} m_1$, m_1 odd.

Euler notes that it suffices to consider an equation $P_n(x) = 0$ for $n = 2^k$, for these equations are the source of the difficulty (if $n \neq 2^k$, then we can find a value of k such that $2^{k-1} < n < 2^k$ and multiply the polynomial $f_n(x)$ by $2^k - n$ factors, say by $x^{2^k - n}$, thus ending up with a polynomial of degree 2^k). He proves the theorem for $n = 4, 8$, and 16 and then goes over to the general case $n = 2^k$. To explain Euler's methods we consider the cases $n = 4$ and $n = 2^k$.

Suppose we are given the equation

$$x^4 + Bx^2 + Cx + D = 0 \tag{3}$$

(we always assume that the equation has been reduced to the form that does not contain x^3). First Euler writes the left side of the equation as the product

$$(x^2 + ux + \lambda)(x^2 - ux + \mu), \tag{4}$$

and using Descartes' method of undetermined coefficients arrives at the equation

$$u^6 + 2Bu^4 + (B^2 - 4D)u^2 - C^2 = 0. \tag{5}$$

According to assumption II, this equation has at least two roots, one of which we take as the value of u. Euler shows that the coefficients λ and μ in the product (4) are rationally expressible in terms of u and the coefficients of the initial equation (3).

Next Euler obtains the same results from general arguments without reliance on computations. He does this in order to extend his assertion to an arbitrary equation of degree 2^k. He makes use of the following (unproved) theorems that later played key roles in the theories of Lagrange and Galois.

A. Every rational symmetric function of the roots of an equation is a rational function of its coefficients (fundamental theorem on symmetric functions).

B. If a rational function $\varphi(x_1, \ldots, x_n)$ of the roots of an equation takes on k different values under all possible permutations of the roots, then those k values satisfy an equation of degree k whose coefficients are rationally expressible in terms of the coefficients of the initial equation.

Euler argued as follows: to any equation of degree n we can "attribute" n roots and write

$$f_n(x) = x^n + a_1 x^{n-1} + \cdots + a_{n-1}x + a_n = (x+\alpha_1)(x+\alpha_2)\cdots(x+\alpha_n), \tag{6}$$

where $\alpha_1, \alpha_2, \cdots, \alpha_n$ are certain symbols with which we can operate as if they were ordinary numbers. Thus the α_i in (6) are the negatives of the roots of f_n and we can write

$$\alpha_1 + \alpha_2 + \cdots + \alpha_n = a_1$$
$$\alpha_1\alpha_2 + \alpha_1\alpha_3 + \cdots + \alpha_{n-1}\alpha_n = a_2, \tag{7}$$
$$\ldots\ldots\ldots\ldots\ldots\ldots\ldots$$
$$\alpha_1\alpha_2\cdots\alpha_n = a_n.$$

The fundamental theorem of algebra asserts that $\alpha_1, \alpha_2, \cdots, \alpha_n$ are real or complex numbers.

Euler proceeds in the same way in the case of $n = 4$. He assumes that equation (3) has roots α, β, γ, and δ. Then u must be the sum of some two of these four roots, i.e., it can take on $\binom{4}{2} = 6$ values under all possible permutations of the roots. From this Euler concludes that u must satisfy an equation of degree 6. He notes that since $\alpha + \beta + \gamma + \delta = 0$, the values of u are

$$u_1 = \alpha + \beta = p \qquad u_4 = \gamma + \delta = -p,$$
$$u_2 = \alpha + \gamma = q \qquad u_5 = \beta + \delta = -q,$$
$$u_3 = \alpha + \delta = r \qquad u_6 = \beta + \gamma = -r,$$

i.e., the required equation for u will be of the form

$$(u^2 - p^2)(u^2 - q^2)(u^2 - r^2) = 0. \qquad (8)$$

This equation is of even degree and its constant term is $-p^2 q^2 r^2$. To be certain that $-p^2 q^2 r^2 < 0$ we must verify that the product pqr is real. To this end Euler shows that

$$pqr = (\alpha + \beta)(\alpha + \gamma)(\alpha + \delta)$$

remains unchanged under all permutations of the roots; this means that it is rationally expressible in terms of the coefficients of equation (3). These general considerations show that u can be chosen as real.[1]

Euler only sketched the proof for the case $n = 2^k$ (theorem 7). He represents the polynomial

$$f_n(x) = x^{2^k} + Bx^{2^k-2} + Cx^{2^k-3} + \cdots \qquad (9)$$

as a product of two factors of degree 2^{k-1} with indeterminate coefficients

$$(x^{2^{k-1}} + ux^{2^{k-1}-1} + \lambda x^{2^{k-1}-2} + \cdots)(x^{2^{k-1}} - ux^{2^{k-1}-1} + \mu x^{2^{k-1}-2} + \cdots).$$

The number of these coefficients is $2^k - 1$ and is thus equal to the number of determining relations.

Since u is the sum of 2^{k-1} of the 2^k roots, the number of different values it can take on is $\binom{2^k}{2^{k-1}} = 2N$, where, as Euler shows, N is odd. From this he concludes that u satisfies an equation of degree $2N$ with real coefficients. It is easy to see that this equation must have the same structure as equation (8), i.e.,

$$(u^2 - p_1^2)(u^2 - p_2^2) \cdots (u^2 - p_N^2) = 0.$$

Euler considers its constant term $-p_1^2 \cdots p_N^2$ and asserts that it is negative (which is not difficult to prove). From this he concludes that u can be chosen to be real. As for the remaining coefficients $\lambda, \mu, \ldots$ he claims that they can be expressed rationally in terms of u and the coefficients $B, C, D, \ldots$ of the polynomial (9).

It is difficult to tell with certainty on what considerations Euler based this conclusion. Lagrange's memoir *On the Form of Imaginary Roots of Equations* (Sur la forme des racines imaginaires des équations, 1772) is a rigorous account of Euler's reduction procedure which fills the gaps in Euler's proofs.

In particular, Lagrange used his theory of similar functions (see §5), developed in his famous memoir *Reflections on Algebraic Solutions of Equations* (Réflexions sur la résolution algébrique des équations, 1771) to prove Euler's assertion about the possibility of expressing $\lambda, \mu, \ldots$ rationally in terms of u and the coefficients of the polynomial (9).

In his proof Lagrange fully accepts Euler's viewpoint: he assumes that one can "attribute" n root symbols to an arbitrary equation of degree n and operate with them according to the usual rules of arithmetic. In particular, one can set down relation (7).

Other 18th-century mathematicians, such as de Foncenex (1759) and Laplace (1795), adopted the same viewpoint in their proofs of the fundamental theorem. They significantly simplified Euler's reduction procedure but regarded his formulation of the issue as completely legitimate.

4. Gauss' criticism

The first to reject Euler's formulation was the young Gauss (1777–1855). His doctoral dissertation (1799) was devoted to the proof of the fundamental theorem. In it Gauss wrote: "Since we cannot imagine forms of magnitudes other than real and imaginary magnitudes $a + b\sqrt{-1}$, it is not entirely clear how what is to be proved differs from what is assumed as a fundamental proposition; but granted one could think of other forms of magnitudes, say $F, F', F'', \ldots$, even then one could not assume without proof that every equation is satisfied either by a real value of x, or by a value of the form $a+b\sqrt{-1}$, or by a value of the form F, or of the form F', and so on. Therefore the fundamental theorem can have only the following sense: every equation can be satisfied either by a real value of the unknown, or by an imaginary of the form $a + b\sqrt{-1}$, or, possibly, by a value of some as yet unknown form, or by a value not representable in any form. How these magnitudes, of which we can form no representation whatever—these shadows of shadows—are to be added or multiplied, this cannot be stated with the kind of clarity required in mathematics" (C. F. Gauss, *Werke*. Göttingen, 1866, Vol. III, pp. 1–2).

In 1815 Gauss returned to the fundamental theorem and this time gave a largely algebraic proof without initially assuming the existence of roots of any form. In Gauss' view such an assumption, "at least at the point when what is involved is a general proof of such a decomposition (of a polynomial into linear and quadratic factors), is simply petitio principii" (ibidem, p. 31).

However, the charge that Euler's proof involved a vicious circle was unfair. And this is best seen by analyzing Gauss' second proof: to avoid the assumption of the existence of root symbols he operates with congruences modulo a certain polynomial, that is, basically, he constructs the splitting field of the initial polynomial.

This method of Gauss was isolated in pure form by Kronecker who gave his famous construction of the splitting field of a polynomial without assuming the existence of the field of complex numbers (1880–1881).

We consider this construction—which Hermann Weyl regarded as one of the first examples of the abstract conception of algebra—in greater detail. The classical treatment of the fundamental theorem was the following: we are given the field $\mathbf{Q}$ of rational numbers (or the field $\mathbf{R}$ of real numbers) and the field $\mathbf{C}$ of complex numbers. One asserts that every polynomial $f(x)$ with coefficients in $\mathbf{Q}$ (or in $\mathbf{R}$) has a root θ in the field $\mathbf{C}$. By adjoining θ to $\mathbf{Q}$ we obtain the simple extension $\mathbf{Q}(\theta)$ in which $f(x) = (x - \theta)g(x)$.

Kronecker gave a new method of construction of the splitting field of a polynomial: Let $f(x)$ be a polynomial of degree n irreducible over a certain field k. We consider the ring of all polynomials $g(x)$ over k and split them into congruence classes mod $f(x)$. We associate with each class a polynomial of degree $\leq n - 1$ as its representative and denote it as a polynomial in θ. If $h(x) \equiv p(x) \pmod{f(x)}$, then we write $h(\theta) = p(\theta)$. In view of the irreducibility of $f(x)$, the classes of polynomials mod $f(x)$ (or the polynomials in θ) form not only a ring but a field $\mathbf{L} = k(\theta)$ with basis $1, \theta, \ldots, \theta^{n-1}$. Since $f(x) \equiv 0 \pmod{f(x)}$, it follows that in this field $f(\theta) = 0$ or $f(x) = (x - \theta)g(x)$.

By iterating this construction we arrive at a field $\mathbf{K}$ in which $f(x)$ splits into linear factors. Then the fundamental theorem of algebra is the assertion that $\mathbf{K}$ is isomorphic to a subfield of the field $\mathbf{C}$ (or that $\mathbf{K}$ can be embedded in $\mathbf{C}$).

But this was Euler's approach, except that he thought that one should postulate the existence of the field $\mathbf{K}$ rather than prove it! This reminds one of his view that it is not necessary to prove that a curve $y = f_{2n-1}(x)$ intersects the x-axis (what counts is not the particular value, $2n - 1$, of the degree of the polynomial that determines the curve but the fact that it is odd). Of course, from a modern viewpoint both assertions require proof but are not instances of a vicious circle!

We conclude by showing how Cauchy (1789–1857), independently of Gauss and Kronecker, used the same approach to construct the field of complex numbers (1847). To this end he singled out the polynomial $x^2 + 1$, which is irreducible over $\mathbf{Q}$ (as well as over $\mathbf{R}$). Then every polynomial is congruent (mod $(x^2 + 1)$) to a linear polynomial $ax + b$ and each class is represented by $a\theta + b$. These classes form a field with the following multiplication rule:

$$(a\theta + b)(c\theta + d) = (ad + bc)\theta + bd + ac\,\theta^2.$$

But $x^2 \equiv -1 (\mod (x^2 + 1))$, so that $\theta^2 = -1$. In other words, this multiplication rule is the same as the one for complex numbers.

It is remarkable that Euler adopted an algebraic viewpoint that was to be rejected at the beginning of the 19th century and that was to be adopted between the 1870s and 1880s.

Thus it was Euler's viewpoint that triumphed in algebra rather than the viewpoint that presupposed the construction of the field of complex numbers followed by a proof of the existence of a root in that field.

We note that, basically, the fundamental theorem of algebra in Euler's formulation coincides with the Weierstrass-Frobenius theorem to the effect that the field of real numbers and the field of complex numbers are the only linear associative and commutative algebras (without zero divisors) over the field of real numbers.[2]

5. The problem of solution of equations by radicals

In addition to the fundamental theorem of algebra, what attracted the attention of 18th-century mathematicians was the problem of solution of equations by radicals. Encouraged by the success of Italian mathematicians in solving cubic and quartic equations they now tried hard to solve the quintic equation. The problem attracted many eminent mathematicians, including Tschirnhaus (1651–1708), Euler, Bézout (1730–1783), Lagrange (1736–1813), and Vandermonde (1735–1796).

In 1683 Tschirnhaus published a paper in *Acta Eruditorum* in which he described a transformation (later named after him) of an equation of degree n

$$x^n + a_1 x^{n-1} + \cdots + a_{n-1}x + a_n = 0$$

by means of a substitution

$$y = b_0 + b_1 x + \cdots + b_{n-1}x^{n-1}$$

into an equation of degree n

$$y^n + c_1 y^{n-1} + \cdots + c_n = 0,$$

whose coefficients $c_1, c_2, \ldots, c_{n-1}$ would vanish for a suitable choice of the undetermined coefficients $b_0, b_1, \ldots, b_{n-1}$. He managed to achieve this for $n = 3$, and thereby obtained a new method for solving a cubic equation. From this he concluded that this can be done for all values of n.

Tschirnhaus had first expounded his method in a letter to Leibniz as early as 1677. Leibniz replied that he thought he could prove that for $n > 4$ the computations required for the determination of the b_i could not be carried out.

While a quintic equation cannot be reduced to a two-term equation by a Tschirnhaus transformation, it can be reduced to an equation of the form

$$x^5 + Ax + B = 0, \tag{10}$$

and the coefficients $b_0, \ldots, b_4$ are determined by an equation of degree at most three. This was shown in 1786 by the Swedish lover of mathematics E. Bring. In 1858 Hermite used the form (10) and proved that the solution of the quintic can be expressed in terms of elliptic modular functions.

Euler tackled the problem of solution of the quintic twice: the first time in 1732–1733, in his memoir *Conjecture on the Form of Roots of Equations of Arbitrary Degree* (Comment. Acad. Petropolitanae, 1738, Vol. 6), and the second time in 1762–1763, in his memoir *On the Solution of Equations of Arbitrary Degree* (Novi Comment. Acad. Petropolitanae, 1762/63, 1764, Vol. 9).

In the first of these papers Euler notes that the solution of equations of degree $n = 2, 3, 4$ reduces to the solution of equations of degree $n = 1, 2, 3$ respectively, and refers to the latter equations as resolvent equations. He assumes that an arbitrary equation

$$x^n = ax^{n-2} + bx^{n-3} + \cdots + q \tag{11}$$

has a resolvent of the form

$$z^{n-1} = \alpha z^{n-2} + \beta z^{n-3} + \cdots \tag{12}$$

If the roots of the latter are $A_1, A_2, \ldots, A_{n-1}$ then, according to Euler, the roots of the original equation are

$$x = \sqrt[n]{A_1} + \sqrt[n]{A_2} + \cdots + \sqrt[n]{A_{n-1}}. \tag{13}$$

Of course, Euler could not have obtained the roots of the quintic in this way. We note also that formula (13) is flawed in that each of its summands can take on n values independently of the values of the remaining summands, so that one can obtain from it n^{n-1} values for x.

In the same paper Euler considers reciprocal equations (i.e., equations $a_0 x^n + a_1 x^{n-1} + \cdots + a_{n-1} x + a_n = 0$ such that $a_i = a_{n-i}$) and proposes for their solution his famous substitution

$$y = x + 1/x.$$

We discuss this in the next chapter.

In the second paper Euler replaces formula (13) by a new formula:

$$x = w + A \sqrt[n]{v} + B \sqrt[n]{v^2} + \cdots + Q \sqrt[n]{v^{n-1}}, \tag{14}$$

where w is real and v satisfies an equation of degree $\leq n - 1$.

This expression, unlike the previous one, takes on exactly n different values, provided the value of $\sqrt[n]{v^2}$ is given by $(\sqrt[n]{v})^2$, and similarly for the other roots. Since it too failed to yield a solution of the general quintic, Euler decided to investigate three special classes of quintic equations, all of which are solvable by radicals. In modern terms, it turns out that the (Galois) groups of the first two classes are cyclic while the (Galois) groups of equations in the third class are dihedral.[3]

What makes the second paper especially important is the generality of the form (14): if equation (11) is solvable by radicals, then a root takes the form (14). This was proved by Abel in his famous proof of the unsolvability of the general quintic by radicals. The form (14) was the starting point of Abel's proof.

After Euler, one of the mathematicians who investigated equations of degree $n \geq 5$ was Bézout. His starting point was that a two-term equation is solvable by radicals. (At the time, this was not an established fact because the roots of unity had not yet been expressed in terms of radicals.) He looked for substitutions that would transform an arbitrary nth degree equation into a two-term equation. Like his predecessors, he found new ways of solving cubic and quartic equations, but for a quintic he obtained "frightful formulas" that led nowhere.

A turning point in the history of the solution of equations by radicals was the appearance in 1770–1771 of Lagrange's famous memoir *Reflections on the Algebraic Solution of Equations* (Réflexions sur la résolution algébrique des équations). It is partly a historico-critical essay. In the first two parts Lagrange analyzes all methods of solution of cubic and quartic equations invented up to his time and shows why none of them is applicable to the general quintic. In the third part he analyzes some classes of solvable equations of higher degree (which we would now characterize as equations with cyclic groups). Finally, in the fourth part, Lagrange makes theoretical deductions based on the whole of the investigated material and concludes that all existing methods reduce to the construction of lower-degree auxiliary equations (resolvents) whose roots are rational functions of the roots of the initial equation. These generalities call for clarification.

Consider the equation

$$f(x) = x^n + a_1 x^{n-1} + \cdots + a_{n-1} x + a_n = 0 \qquad (15)$$

with literal coefficients $a_1, \ldots, a_n$ and roots $x_1, x_2, \ldots, x_n$. Let $y = \varphi(x_1, \ldots, x_n)$ be a rational function of the roots. If this function is invariant under all permutations of the roots (equivalently, under the action of the symmetric group S_n) then, as Lagrange shows, it is rationally expressible in terms of the coefficients of equation (15); this is Euler's theorem A. Now suppose that under all permutations of the roots $y = \varphi(x_1, \ldots, x_n)$ takes on k different values $y_1, y_2, \ldots, y_k$. Lagrange shows that in that case y satisfies an equation of degree k whose coefficients are rationally expressible in terms of the coefficients of the initial equation (Euler's theorem B.). Indeed, consider the expression

$$(y - y_1) \cdots (y - y_k) = y^k - (y_1 + \cdots + y_k) y^{k-1} + \cdots + (-1)^k y_1 \cdots y_k = 0.$$

Since, as is easy to see, its coefficients are symmetric functions of the roots of equation (15), they are rationally expressible in terms of $a_1, \ldots, a_n$. In general, $k = n!$, but sometimes one can find functions $y = \varphi(x_1, \ldots, x_n)$ that take on $k < n$ values under the action of S_n. Thus, for example, regardless of the value of n, the function

$$\prod_{\substack{i,j=1 \\ i<j}}^{n} (x_i - x_j)$$

takes on just two values.

Our argument shows that the degree k of the auxiliary equation depends not on the form of the function $y = \varphi(x_1, \ldots, x_n)$ but only on the number of values it takes on under the action of S_n. If H is the set of elements of S_n that leave φ invariant, then H is a subgroup of S_n (the identity of S_n is in H; if $h, \sigma \in H$ then $h\sigma \in H$; finally, if $h \in H$, then $h^{-1} \in H$) and the number of different values taken on by φ is equal to the index of H in

S_n. The latter follows from the coset decomposition of S_n with respect to the subgroup H:

$$S_n = H + \sigma_1 H + \cdots + \sigma_{k-1} H, \qquad \sigma_i \notin H, \ i = 1, \ldots, k - 1.$$

Lagrange obtained this theorem in a similar way. It implies that the order of a subgroup is a divisor of the order of the group, a result later named after Lagrange.

Lagrange calls two functions $\varphi(x_1, \ldots, x_n)$ and $\psi(x_1, \ldots, x_n)$ of the roots of an equation similar if all permutations of the roots that leave φ invariant also leave ψ invariant and conversely. Of course, this set H of "joint symmetries" of φ and ψ is a subgroup of S_n (we say that the two functions belong to the same subgroup H of S_n). He shows that similar functions are rationally expressible in terms of one another and of the coefficients of the equation. In the language of field theory this means that two functions that belong to the same subgroup lie in the same field. Later, this correspondence between the subgroups of the group of permutations of the roots of an equation and the subfields of the splitting field of a polynomial $f(x) : \mathbf{K} = k(x_1, \ldots, x_n)$ occupied a key position in the investigations of Gauss and in the theory of Galois.

Lagrange's analysis led him to the following conclusion: "These observations give, if I am not mistaken, the true principles of the solution of equations...all is reduced, as we see, to a kind of calculus of combinations, by which one finds a priori the results that one might expect" (J. L. Lagrange, *Oeuvres*: Vol. 1–14, Paris, 1867–1892. Vol. 4, p. 403).

Thus Lagrange must be given credit for the idea that the solution of an equation by radicals depends on the group of permutations of its roots and on the subgroups of that group. The idea of "similar" functions brought Lagrange to the consideration of the simplest functions of the roots, for example,

$$t = x_1 + \alpha x_2 + \alpha^2 x_3 + \cdots + \alpha^{n-1} x_n,$$

where $x_1, \ldots, x_n$ are the roots of the initial equation and $\alpha^n = 1, \alpha \neq 1$. The function t is called the Lagrange resolvent. It is easy to see that $\theta = t^n$ is invariant under the action of a subgroup H of cyclic permutations. If the group of an equation is a cyclic subgroup of permutations (as in the case of $x^n - 1 = 0$, or of $x^n - A = 0$), then the equation can be solved by means of this resolvent. But if the order of the group of an equation is $n!$, then θ will take on $n!/n = (n-1)!$ different values. For $n = 3$ the number of different values is two, which is why a cubic equation is solvable by radicals. But if

$n = 5$, then θ will take on $5!/5 = 4! = 24$ different values, i.e., it will satisfy an equation of degree 24.

Next Lagrange considered a more extensive group of permutations of the roots of an equation, namely permutations of the form $x_k \rightarrow x_{ak+b}$, where b is an arbitrary integer, a is an integer relatively prime to n, and $ak + b$ is taken mod n. The order of this group is $n(n - 1)$ if n is a prime and $n\varphi(n)$ otherwise; here $\varphi(n)$ is the Euler function (it denotes the number of positive integers less than n and relatively prime to n). When n is a prime, this is the one-dimensional linear affine group modulo n, which generalizes the case $n = 5$ mentioned earlier. Under the action of this group, θ takes on $n - 1$ different values. This means that any symmetric function of $\theta_1, \theta_2, \ldots, \theta_{n-1}$ is invariant under the permutations of this group. Let ψ be such a function. Under the action of S_n, ψ will take on $n!/n(n - 1) = (n - 2)!$ different values, i.e., it will satisfy an equation of degree $(n - 2)!$.

This means that a cubic can be reduced to a linear equation and a quintic to an equation of degree six, i.e., in the general case of a quintic we are not led to a solution by radicals.

Lagrange arrived at the conclusion that "It is very doubtful that the methods just discussed can give a complete solution of the quintic equation, and, all the more so, of equations of higher degrees" (ibidem, Vol. 5, p. 355).

We now use the Lagrange resolvent to give a solution of the cubic equation

$$y^3 + py + q.$$

Let y_1, y_2, y_3 be the roots of this equation ($y_1 + y_2 + y_3 = 0$). Let

$$t = y_1 + \alpha y_2 + \alpha^2 y_3, \qquad \text{where} \quad \alpha^3 = 1, \ \alpha \neq 1,$$

and let

$$\theta = t^3 = (y_1 + \alpha y_2 + \alpha^2 y_3)^3.$$

Under the cyclic permutation $y_k \rightarrow y_{k+1}$, t is multiplied by α^2, so that θ remains unchanged. The cyclic subgroup H of S_3, generated by $y_k \rightarrow y_{k+1}$, has index 2. Hence θ takes on two different values:

$$\theta_1 = (y_1 + \alpha y_2 + \alpha^2 y_3)^3; \qquad \theta_2 = (y_1 + \alpha y_3 + \alpha^2 y_2)^3.$$

Now $\theta_1 \theta_2 = -27p^3$ and $\theta_1 + \theta_2 = 27q$. This means that θ_1 and θ_2 can be determined from a quadratic equation with rational coefficients. Since

$$y_1 + \alpha y_2 + \alpha^2 y_3 = \sqrt[3]{\theta_1},$$

$$y_1 + \alpha y_3 + \alpha^2 y_2 = \sqrt[3]{\theta_2},$$

$$y_1 + y_2 + y_3 = 0,$$

it follows that

$$3y_1 = \sqrt[3]{\theta_1} + \sqrt[3]{\theta_2}; \quad 3y_2 = \alpha^2 \sqrt[3]{\theta_1} + \alpha \sqrt[3]{\theta_2}; \quad 3y_3 = \alpha \sqrt[3]{\theta_1} + \alpha^2 \sqrt[3]{\theta_2}.$$

In his *Memoir on the Solution of Equations* Vandermonde arrived at analogous (though less general) conclusions at almost the same time as Lagrange. In this paper he also tried to express the roots of $x^n - 1 = 0$ by radicals but was unable to go beyond the value $n = 11$. His very interesting findings had no effect on the evolution of algebra. His paper was "noticed" and commented on only in the 20th century.

6. Proof of the unsolvability of the general quintic by radicals

Thus, according to Lagrange, the problem of solving an equation by radicals requires knowledge of the subgroups of the group G of permutations of its roots. If the degree of the equation is n and G contains a subgroup of index $k < n$, then the solution of the equation can be reduced to the solution of an equation of degree k.

Lagrange could not claim that the general quintic is definitely not solvable by radicals because 1) he did not know that the symmetric group S_5 has no subgroups of index between 2 and 5, and because 2) he had no proof of the fact that every intermediate radical in the solution of an equation is a rational function of its roots.

After Lagrange, investigations of the problem of solvability of an equation by radicals followed one of two routes:

1. Investigation of equations with arbitrary literal coefficients whose (Galois) group is the symmetric group S_n.

2. Investigation of equations with numerical coefficients, finding classes of equations of arbitrary degree that are solvable by radicals, and finding equations with prescribed coefficients that are solvable by radicals.

Paolo Ruffini (1765–1822) and Niels Henrik Abel (1802–1829) followed the first of these routes. In 1799 Ruffini published a proof of unsolvability by radicals of the general quintic. He used permutations

$$\begin{pmatrix} 1 & 2 & 3 & 4 & 5 \\ i_1 & i_2 & i_3 & i_4 & i_5 \end{pmatrix}$$

as well as products of permutations. In his proof Ruffini assumed that all intermediate radicals in an expression for a root of an equation are rational functions of the roots of the initial equation. This was a gap he was unable to fill.

When he was still a student, Abel produced a "proof" of the solvability by radicals of the general quintic. He spotted the error himself. In 1824 he published a short proof of the unsolvability by radicals of the general quintic and in 1824 a complete proof. This proof appeared in the first issue of *Crelle's Journal* (Journal für die reine und angewandte Mathematik). Abel proved that if a quintic equation is solvable by radicals, then its roots have the form found by Euler (see §5, Chapter V, formula (14)). He also showed that all intermediate radicals are rational functions of the roots of the initial equation.[4] In his proof Abel used Cauchy's theorem that for $n > 5$, S_n has no subgroups of index > 2 and $< p$, where p is the largest prime not exceeding n. The group S_n always has a subgroup of index 2, namely the alternating subgroup A_n of even permutations. That is why one can always find a function of the roots that takes on just two values:

$$\prod_{\substack{i,j=1 \\ i<j}}^{n} (x_i - x_j).$$

In the next chapter we discuss the second route, far more complex than the first and rich in new ideas.

Editor's notes

[1] There is some awkwardness here because the meaning of $x_1, \ldots, x_n$ in Theorems A and B above is not clear. As stated, these theorems are valid when $x_1, \ldots, x_n$ are independent variables. Yet in Euler's argument for $n = 4$, α, β, γ and δ are roots of an equation with real coefficients. To see how this causes problems, note that if we regard α, β, γ and δ as variables, then the product $pqr = (\alpha + \beta)(\alpha + \gamma)(\alpha + \delta)$ is not invariant under all permutations, yet it is invariant if α, β, γ and δ are regarded as roots of equation (3). Gauss was careful to distinguish between these two cases in his second proof of the fundamental theorem of algebra, though the complete resolution of this confusion came only after the emergence of Galois theory.

[2] The following may be of use. In Gauss' second proof (late 1815) of the Fundamental Theorem of Algebra, the congruences modulo polynomials are only implicit in the argument. In 1818 such congruences appear more explicitly in Gauss' sixth proof of Quadratic Reciprocity. This proof is a modification

of an earlier proof which uses Gauss sums similar to the periods defined in 1 of Chapter VII. In the sixth proof, he replaces Gauss sums with the polynomial $x - x^g + \cdots - x^{g^{n-2}}$, where n is an odd prime and g is a primitive root modulo $n - 1$. Gauss considers this and other polynomials modulo $X = x^{n-1} + \cdots + x + 1$, which in effect means that x plays the role of a primitive n-th root of unity. As explained above, Cauchy used this idea in 1847 when he represented $a + bi$ as $a + bx$ modulo $x^2 + 1$, and the culmination of this line of thought came with Kronecker's work in 1880–1881. Kronecker mentions Gauss' 1815 proof as a source of inspiration.

[3] By Galois theory, the (Galois) group of an irreducible quintic is one of five groups: S_5, A_5, the cyclic group of order 5, the dihedral group of order 10, or the group of order 20 given by the transformations $k \rightarrow ak + b$, where a, b, k are integers, a is relatively prime to 5, and k and $ak + b$ are taken modulo 5. This is the *one-dimensional affine linear group modulo 5*. The last three groups are solvable.

[4] Abel's proof of this assertion is not complete. The article "On the nonsolvability of the general polynomial" by R. Ayoub (*Amer. Math. Monthly,* vol. 89, 1982, 397–401) gives a modern version of Abel's argument and explains (on p. 399) where the gap occurs.

CHAPTER 7

The theory of algebraic equations in the 19th century

1. Cyclotomic equations

In the 19th century, algebra, like other areas of mathematics (such as geometry and analysis), underwent a radical transformation. One type of algebraic thinking was replaced by another. A turning point in the evolution of algebra was Galois theory whose prototype was Gauss' theory of cyclotomic equations. This theory served as a model for the investigations of Abel, Galois, and other 19th-century algebraists.

While still a student in Göttingen young Gauss began to study the problem of ruler and compass construction of regular polygons. The problem had a long history. The geometers of antiquity had constructed regular n-gons for $n = 3, 3 \cdot 2^k$, 4, $4 \cdot 2^k$, 5, and $5 \cdot 2^k$. What remained an open problem was the possibility of ruler and compass construction of a regular 7-gon or a regular 11-gon.

Gauss reduced the problem of ruler and compass construction of regular n-gons to the problem of solution of the equation

$$X = \frac{x^n - 1}{x - 1} = x^{n-1} + x^{n-2} + \cdots + x + 1 = 0, \qquad (1)$$

whose roots $x_k = e^{\frac{2\pi i}{n} k} = \cos \frac{2\pi k}{n} + i \sin \frac{2\pi k}{n}$ are located at the vertices of a regular n-gon. Equation (1) came to be known as the cyclotomic equation.

By investigating this equation Gauss discovered that for $n = 17$ its roots can be expressed in terms of quadratic radicals and can therefore be constructed by ruler and compass. As a result of this discovery Gauss, who until then was torn between philology and mathematics, decided to study mathematics. From then on he kept a "Mathematical diary". The first entry in this diary, dated 30 March, 1796, was: "The principles underlying the division of a circle, namely its geometric division into seventeen parts."

109

Gauss' diary contains 146 entries. The dates of 102 entries fall between 1796 and 1800. Thus, as in the case of most mathematicians, his fundamental ideas came to him in his youth.

Gauss presented a complete account of the theory of cyclotomic equations in the last (the seventh) chapter of his famous *Arithmetical Investigations* (Disquisitiones arithmeticae) of 1801. This work immediately elevated him to the level of a leading mathematician of his time. He was soon to be referred to as "princeps mathematicorum".

Although Gauss had results for other types of of transcendental functions (such as the lemniscatic functions), he limited his discussion in *Disquisitiones* to the case of circular functions "both for the sake of brevity and in order that the new principles of this theory may be more easily understood" (Gauss, *Disquisitiones*, Art. 335, p. 407 of A. A. Clarke's translation, Springer, 1986). This implies that Gauss knew the solution of the general case as well. We will return to this matter later.

In his study of the equation $X = 0$ Gauss assumes that n is prime. He shows that for such n equation (1) is irreducible, i.e., cannot be written as a product of two factors with rational integral coefficients. He describes his subsequent objective as follows: "We will show that if the number $n - 1$ is resolved in any way into integral factors α, β, γ, etc. (we can assume each of them is prime), X can be resolved into α factors of degree $(n - 1)/\alpha$ with coefficients determined by an equation of degree α; each of these will be resolved into β others of degree $(n - 1)/\alpha\beta$ with the aid of an equation of degree β, etc." (ibidem, Art. 342, p. 414). Thus if $n - 1 = \alpha\beta \cdots \nu$, then the solution of the equation $X = 0$ is reducible to the successive solution of equations of degree $\alpha, \beta, \ldots, \nu$. It follows that a regular n-gon can be constructed by ruler and compass if $n - 1 = 2^k$, or $n = 2^k + 1$. But $2^k + 1$ can be a prime only if $k = 2^S$. This implies that if a prime p is of the form $2^{2^S} + 1$, then the corresponding p-gon can be constructed by ruler and compass. For $S = 0$ and 1 we get $p = 3$ and 5, i.e., the regular polygons constructed by the Pythagoreans. For $S = 2$ and 3 we get $p = 17$ and 257, i.e., the next two regular polygons (with a prime number of angles) constructible by ruler and compass.

Numbers of the form $2^{2^S} + 1$ are called Fermat numbers. Fermat thought that all such numbers are primes but Euler showed that for $S = 5$ we get a composite number. The question of whether the number of Fermat primes is finite or infinite remains open.

What were Gauss' "completely novel principles"? To answer this question we must look at the way cyclotomic equations were solved before Gauss.

We consider Euler's method of solution of equation (1) for $n = 7$:

$$(x^7 - 1)/(x - 1) = x^6 + x^5 + \cdots + 1 = 0. \tag{2}$$

If we divide this equation by x^3 and make the substitution

$$y = x + 1/x, \tag{3}$$

then we obtain the equation

$$y^3 + y^2 - 2y - 1 = 0. \tag{4}$$

This equation has three roots θ_1, θ_2, and θ_3. Substituting them successively in (3) we obtain three quadratic equations:

$$x^2 - \theta_i x + 1, \quad i = 1, 2, 3. \tag{3'}$$

By solving these equations we obtain the roots of equation (3). In his memoir of 1771 (referred to in the previous chapter) Vandermonde used this method to solve equation (1) for the value $n = 11$ but was unable to go beyond it. To go further one had to devise a method of constructing functions of the roots of equation (1) that would, as a result of all permutations of these roots, take on just α different values for each α that divides $n - 1$.

We will now try to obtain Euler's results from general considerations. The roots of $x^7 - 1 = 0$ can be written as $r, r^2, \ldots, r^6, r^7 = 1$. We have $n - 1 = 6 = 2 \cdot 3$. The substitution (3) can be written as

$$y = r + 1/r = r + r^6.$$

The function y is invariant under the permutations of the subgroup $H = \{r \to r, r \to r^6\}$ and takes on the three values

$$\theta_1 = r + r^6, \theta_2 = r^2 + r^5, \theta_3 = r^3 + r^4$$

under the permutations of the group $G = \{r \to r^k,\ k = 1, \ldots, 6\}$. By a Galois-theory generalization of Lagrange's theorem, θ_1, θ_2, and θ_3 are roots of a cubic equation with rational coefficients. The group of permutations of this equation is G/H. To the chain of groups $G \supset H \supset E$ there corresponds the chain of fields

$$\mathbf{Q} \subset \mathbf{L} \subset \mathbf{K},$$

where $\mathbf{L}$ has degree 3 over $\mathbf{Q}$, and $\mathbf{K}$, the splitting field of equation (2), has degree 2 over $\mathbf{L}$.

We see that the problem reduces to finding a rational function of the roots of equation (2) that takes on just three values under all permutations of these roots.

But how is one to proceed in the general case? How is one to construct rational functions of the roots of equation (1) that would take on just α different values for any divisor α of $n - 1$?

Here is what Gauss did. His first step was to rewrite the set $\Omega = \{r, r^2, \ldots, r^{n-1}\}$ of roots of equation (1). Specifically, he picked a primitive root $g \bmod n$ (i.e., $g^{n-1} \equiv 1 \pmod{n}$ but $g^\lambda \not\equiv 1 \pmod{n}$ if $\lambda < n - 1$; the existence of a primitive root mod any prime was first proved by Gauss) and wrote Ω as

$$r^{\lambda g}, \ r^{\lambda g^2}, \ldots, \ r^{\lambda g^{n-1}}, \tag{5}$$

where λ is any integer not divisible by n. Indeed, as Gauss notes, $r^{\lambda g^\mu}$ and $r^{\lambda g^\nu}$ "will be identical or different according as μ and ν are congruent or noncongruent modulo $n - 1$" (ibidem, Art. 343, p. 415). Thus, apart from order, the sequence (5) represents the set Ω.

The representation (5) shows that the group G of permutations of the roots of equation (1) (or the Galois group of this equation) is cyclic of order $n - 1$ and its generating permutation is $r \to r^g$. Furthermore, this representation establishes an isomorphism between G and the group of residue classes mod $(n - 1)$.

If $n - 1 = ef$, then Gauss constructs the function

$$(f, \lambda) = r^\lambda + r^{\lambda g^e} + r^{\lambda g^{2e}} + \cdots + r^{\lambda g^{e(f-1)}}, \quad (\lambda, n) = 1, \tag{6}$$

of the roots that takes on exactly e different values under the action of G. This function, which Gauss called an f-term period, is invariant under the action of the subgroup H of order f generated by the permutation $r \to r^{g^e}$. Essentially, this shows that for every divisor f of $n - 1$, the order of the cyclic group G, there is a cyclic subgroup of order f. Under the action of G the period (6) takes on exactly e different values:

$$(f, 1), \ (f, g), \ (f, g^2), \ldots, (f, g^{e-1}),$$

where $(f, g^\mu) = r^{g^\mu} + r^{g^{\mu+e}} + \cdots + r^{g^{\mu+e(f-1)}}$. By Lagrange's theorem, the f-term periods are roots of an equation of degree e with rational coefficients:

$$\varphi_e(y) = 0. \tag{7}$$

It is easy to see that the Galois group of this equation is the cyclic group G/H. To find the roots of equation (1) one must solve e equations of degree f whose coefficients are rationally expressible in terms of the roots of equation (7).

Gauss shows that the product of two f-term periods is expressible as the sum of f-term periods, and that if $p = (f, \lambda)$ is an f-term period with $(\lambda, n) = 1$, then all remaining f-term periods can be represented in the form

$$a + bp + cp^2 + \cdots + mp^{e-1},$$

where $a, b, c, \ldots, m$ are rational; in other words, the f-term period p generates a field **L** which is normal (this term is explained in the penultimate paragraph of the next section; of course, Gauss did not use these terms). It is easy to see that

$$\mathbf{Q} \subset \mathbf{L} \subset \mathbf{K},$$

where **K** is the splitting field of equation (1). The degree of **L** over **Q** is e and the degree of **K** over **L** is f.

Gauss takes for e the smallest prime divisor of $n - 1$, call it α, and obtains periods of maximal length $f = (n - 1)/\alpha$. Then he factors f. If β is the least prime divisor of f, then every f-term period can be written as a sum of β periods of length f/β, and so on. We reproduce Gauss' table of periods for $n = 19$:[1]

$$
(18,1)
\begin{cases}
(6,1) \begin{cases} (2,1) \ldots [1],[18] \\ (2,8) \ldots [8],[11] \\ (2,7) \ldots [7],[12] \end{cases} \\[2em]
(6,2) \begin{cases} (2,2) \ldots [2],[17] \\ (2,16) \ldots [3],[16] \\ (2,14) \ldots [5],[14] \end{cases} \\[2em]
(6,3) \begin{cases} (2,4) \ldots [4],[15] \\ (2,13) \ldots [6],[13] \\ (2,9) \ldots [9],[10] \end{cases}
\end{cases}
$$

This procedure yields a sequence

$$G \supset H_1 \supset H_2 \supset \cdots \supset H_\nu \supset E$$

of subgroups of G. Their orders are $(n - 1)/\alpha, (n - 1)/(\alpha\beta), \ldots, \tau$ respectively, and the orders of the factor groups $G/H_1, H_1/H_2, \ldots, H_\nu$ are $\alpha, \beta, \ldots, \tau$ respectively. The field **K** can be obtained from **Q** by a sequence of extensions

$$\mathbf{Q} \subset \mathbf{L}_1 \subset \cdots \subset \mathbf{L}_\nu = \mathbf{K},$$

where the degree of $\mathbf{L}_1$ over $\mathbf{Q}$ is α, that of $\mathbf{L}_2$ over $\mathbf{L}_1$ is β, and so on. Gauss shows that all intermediate equations $\varphi_\alpha = 0, \varphi_\beta = 0, \ldots, \varphi_\nu = 0$ are solvable by radicals. Their (Galois) groups are cyclic of prime order. Gauss obtains the solutions of these equations by means of Lagrange resolvents.

Later, the kind of correspondence in which the subgroups of a group G are associated with the subfields of a field $\mathbf{K}$ became fundamental in Galois theory, a theory in which the study of the structure of an infinite field $\mathbf{K}$, the splitting field of a polynomial $f_n(x)$, is reduced to the study of the structure of a finite group G, the group of permutations of the roots of the equation.

As pointed out earlier, Gauss' theory implies that a sufficient condition for a polygon with a prime number n of sides to be constructible with ruler and compass is that $n - 1 = 2^k$. At the end of his book Gauss claims that this condition is also necessary and that he proved its necessity. But, as he puts it, "The limits of the present work exclude this demonstration here" (ibidem, Art. 365, p. 459).

Thus for an equation with a cyclic group of permutations of the roots Gauss constructed a theory that later became a model for the investigations of Abel and Galois.

2. Equations with an Abelian group

Two remarks made by Gauss in his *Arithmetical Investigations* justify the conclusion that he could solve equations not only with a cyclic (Galois) group but with an arbitrary commutative group. At the beginning of the VIIth part of this book (discussed in the previous section) Gauss writes: "The principles of the theory which we are about to present extend far beyond the point to which they are developed here. Indeed, they are applicable not only to cyclotomic but also to many transcendental functions, for example, to functions that depend on the integral $\int dx/\sqrt{1 - x^4}$." Here Gauss has in mind the division of a lemniscate into n equal parts. This problem reduces to an equation of degree n^2, the permutations of whose roots form a commutative group. Gauss says that he intends to investigate this problem in a special essay he is working on. This essay was never published, and what has been preserved is Gauss' rough notes.

Further, Gauss writes that cyclotomic functions can also be "considered in full generality", but as mentioned in §1 he limits his account to the case of division of a circle into a prime number of equal parts. In case of a composite n the group of the equation is commutative but not necessarily cyclic. Thus we can be certain that Gauss extended his methods to the case of equations with a commutative but not necessarily cyclic group. This is confirmed by many entries in his diary recently analyzed by O. Neumann.

Since Gauss never published these insights, it is Abel who must be given credit for the discovery of a large group of equations solvable by radicals. In 1829 *Crelle's Journal* published his *Memoir on a Certain Class of Algebraically Solvable Equations* (Mémoire sur une classe particulière d'équations résolubles algébriquement). It is possible that the problem which motivated the investigations of both Gauss and Abel was the problem of division of a lemniscate into n equal parts.[2]

In his paper Abel explicitly introduced the concept of a domain of rationality (i.e., of the ground field). He defined the domain of rationality with respect to quantities $a_1, \ldots, a_n$ to be the set of all quantities obtained from $a_1, \ldots, a_n$ and from the real (or rational) numbers by applying to them the four arithmetical operations. The introduction of this concept was essential for all general investigations in the theory of equations.

Abel's second important step was his proof of the solvability by radicals of a remarkable class of algebraic equations defined by him by the following two conditions.

1. All roots of an equation in this class are rational functions of one of them: $x_k = \theta_k(x_1)$. (We now call an equation with this property *normal*).

2. If $\theta_k(x_1)$ and $\theta_l(x_1)$ are any two roots, then $\theta_k(\theta_l(x_1)) = \theta_l(\theta_k(x_1))$.

The class of equations singled out by Abel is now called the class of normal equations with Abelian Galois group.[3]

His manuscript shows that in the last year of his life Abel tried to find a general criterion for the solvability by radicals of an arbitrary (polynomial) equation with prescribed numerical coefficients. Already in 1826, in a letter to his teacher Holmboe dated 16 January, he wrote: "I am trying to solve the following problem: find all algebraically solvable equations. I have not yet reached this aim, but I gather that things will go well: in the case of equations of prime degree there are no major difficulties, but if this number is composite then there is hell to pay. I made an application to the quintic and succeeded in solving the problem in that case." Later Abel obtained many important theorems which showed that he was on the right track, but his premature death prevented him from completing the work.

3. Galois theory

The problem of algebraic solution of equations was solved completely by the brilliant French mathematician Evariste Galois (1811–1832). His ideas were not appreciated during his short life. Initially the memoirs he submitted to the Academy of Sciences were mislaid in the papers of Cauchy and Fourier. Finally, they were read by Lacroix and Poisson, neither of whom understood them.

Galois was killed in a duel on 30 May 1832, a few days after his release from prison where he ended up as one of the organizers of the Bastille Day demonstration that took place on 14 July 1831. The night before the duel he wrote a letter to his friend Auguste Chevalier in which he set down his fundamental results. In addition to the theory of equations (discussed below) they dealt with the general theory of algebraic functions. In this area too Galois went far beyond his contemporaries. The well-known French mathematician Émile Picard wrote at the end of the 19th century: "There arises the conviction that he [Galois] knew the most essential results of the theory of Abelian integrals obtained by Riemann 25 years later. We are astounded to see that Galois speaks of periods of Abelian integrals.... The theorems are correct" (*Oeuvres mathématiques d'Evariste Galois*, ed. J. Picard, Paris, 1897).

Toward the end of his letter to Chevalier Galois wrote: "I have often in my life dared to advance propositions that I was not sure of; but everything I have written here has been in my head for over a year, and it is too much in my interest not to deceive myself so that someone could suspect me of stating theorems of which I did not have a complete proof." (What makes these words all the more remarkable is that the proofs of the theorems in question use the theory of functions of a complex variable in the form given to it by Riemann and Weierstrass.)

The letter concludes with a plea to his friend: "You will publicly beg Jacobi or Gauss to give their opinion not of the truth but of the importance of the theorems." (Galois, "Letter to Auguste Chevalier", in H. Wussing, *The Genesis of the Abstract Group Concept*, MIT Press, 1984, p. 116)

Galois' fundamental results on the theory of equations are contained in his *Memoir on the Conditions for the Solvability of Equations by Radicals* (completed in January 1831 and published in 1846). This paper was the basis of a new style of thinking in algebra. In the words of Kolmogorov, the algorithm of formulas was replaced in the 19th century by the algorithm of concepts. Galois' memoir is a perfect illustration of this thought.

It is not our intention to give an account of Galois theory. What we will try to do is give the reader an appreciation of its fundamental ideas. We will often use the language of modern mathematics.

The problem is to decide whether an equation

$$f(x) = x^n + a_1 x^{n-1} + \cdots + a_{n-1}x + a_n = 0, \qquad (7)$$

with given numerical coefficients $a_1, a_2, \ldots, a_n$, is solvable by radicals.

Galois begins his paper with the introduction of the new concepts of *field* and *group*. His name for field is *domain of rationality*. He stresses its major importance. He emphasizes that the notions of *reducibility* and *irreducibility* of an equation make sense only relative to a given *domain of rationality*. His definition follows:

"One can agree to regard as rational all rational functions of a certain number of determined quantities, supposed to be known *a priori*. For example, one can choose a particular root of a whole number and regard as rational all rational functions of this radical. . . .

With these conventions, we shall call *rational* any quantity which can be expressed as a rational function of the coefficients of the equation and of a certain number of *adjoined* quantities arbitrarily agreed upon." (Galois, *Memoir*. . ., in H. M. Edwards, *Galois Theory*, Springer, 1984, p. 101).

The adjoint quantities influence the properties of an equation and the difficulties associated with it. Thus the equation $X = \frac{x^p - 1}{x - 1} = 0$, p prime, is irreducible over the field $\mathbf{Q}$ of rational numbers, but if one adjoins to $\mathbf{Q}$ "a root of one of Gauss' auxiliary equations, this equation decomposes into factors" (ibidem, p. 102).

Then Galois introduces the notion of a *group of substitutions* (the term "group" is due to Galois): "Substitutions are the passage from one permutation to another" (ibidem, p. 102).

He notes that a substitution is independent of the disposition of the letters in the initial permutation, and states the fundamental fact that if substitutions S and T are in G then so is ST. Then Galois begins to develop his theory. As the domain of rationality he takes $\mathbf{Q}(a_1, \ldots, a_n) = \mathbf{Q}_0$, where $\mathbf{Q}$ is the field of rational numbers.

Galois usually adjoins to $\mathbf{Q}_0$ all necessary roots of unity. Thus the equation

$$x^p - a = 0$$

becomes solvable by radicals if we adjoin to the domain of rationality one radical $\sqrt[p]{a}$.

All equations considered up to the time of Galois were normal (see condition 1 above) and most had a solvable Galois group.

To obtain a normal equation Galois goes over from equation (7) to a certain new equation $F_g(y) = 0$. Specifically, he considers a rational function

$$\theta = c_1 x_1 + c_2 x_2 + \cdots + c_n x_n \tag{8}$$

of the roots of equation (7) that takes on $n!$ different values under all possible permutations of these roots. The function θ satisfies an equation $F_m(y) = 0$, $m = n!$, whose coefficients are rationally expressible in terms of the coefficients of the initial equation (7). If the polynomial $F_m(y)$ is reducible, then one takes its irreducible factor of which θ is a root. Denote this factor by $F_g(y)$. The equation

$$F_g(y) = y^g + b_1 y^{g-1} + \cdots + b_g = 0 \tag{9}$$

is called a Galois resolvent.

Galois shows that equation (9) is normal—i.e., its roots θ_k satisfy equations of the form $\theta_k = r_k(\theta)$, where θ stands for θ_1 and $r_k(\theta)$ is some rational function—and that all roots of equation (7) are rationally expressible in terms of θ: $x_k = \varphi_k(\theta)$.

In other words, the splitting field $\mathbf{K} = \mathbf{Q}_0(x_1, \ldots, x_n)$ of equation (7) can be obtained by the adjunction of a single element θ: $\mathbf{K} = \mathbf{Q}_0(\theta)$. The element θ is called a primitive element of the field $\mathbf{K}$.

The question of solvability of equation (7) reduces to the study of the structure of the field $\mathbf{K}$. Can this field be obtained by successive adjunction of radicals to $\mathbf{Q}_0$? Can we construct a chain of fields $\mathbf{L}_1, \ldots, \mathbf{L}_S$ such that

$$\mathbf{L}_1 \subseteq \mathbf{Q}_0 \left(\sqrt[p_1]{c} \right), \quad c \in \mathbf{Q}_0,$$

$$\mathbf{L}_2 \subseteq \mathbf{L}_1 \left(\sqrt[p_2]{c_1} \right), \quad c_1 \in \mathbf{L}_1,$$

$$\ldots\ldots\ldots\ldots\ldots\ldots\ldots\ldots\ldots\ldots\ldots\ldots\ldots$$

$$\mathbf{K} \subseteq \mathbf{L}_S = \mathbf{L}_{S-1} \left(\sqrt[p_S]{c_{S-1}} \right), \quad c_{S-1} \in \mathbf{L}_{S-1}?$$

Galois reduced the problem to the study of the structure of a finite group, the group G of automorphisms of the field $\mathbf{K}$ that leave the elements of the groundfield $\mathbf{Q}_0$ fixed. (We are using modern terminology. Galois defined G as the set of permutations of the roots of equation (1) that leave unaltered the rational relations between the roots.) Since equation (9) is normal, this group has order g and consists of the permutations

$$\theta_1 \to \theta_1, \ \theta_1 \to \theta_2, \ldots, \ \theta_1 \to \theta_g.$$

In general, this group is noncommutative (if there are no nontrivial relations between the roots then $G = S_n$). It is known as the Galois group. If

H is a subgroup of G, then, clearly, the elements of $\mathbf{K}$ invariant under the permutations in H form a subfield $\mathbf{L}$. In general, this subfield is not normal (i.e., it is not determined by a normal equation). Galois realized that in order to obtain normal equations (and normal subfields associated with them) one must take only subgroups H such that the cosets $H, g_1 H, \ldots, g_{S-1} H$ form a group, now known as the factor group G/H. In other words, H must be a normal subgroup of G. In his memoir Galois defined a normal subgroup by this property but in the letter to Chevalier he defined it by the coincidence of the two coset decompositions of G with respect to H:

$$G = H + Hs + Hs' + \cdots;$$

$$G = H + \tau H + \tau' H + \cdots.$$

He wrote: "Usually these two kinds of decomposition do not coincide. When they do, then we say that the decomposition is *proper*. ... It is easy to see that if a group of an equation does not admit any proper decomposition then one can transform it all one wants; the groups of the transformed equations will always have the same number of permutations" (Galois, "Letter to Auguste Chevalier", in H. Wussing, *The Genesis of the Abstract Group Concept*, MIT Press, 1984, p. 115)

Thus Galois introduced into group theory two new extremely important concepts, namely those of a normal subgroup and of a factor group.

If H is a normal subgroup of order h and index p and $\mathbf{L}$ is the subfield of $\mathbf{K}$ whose elements are invariant under H, then $\mathbf{K} \supset \mathbf{L} \supset \mathbf{Q}_0$ and, as Galois shows, the degree of $\mathbf{L}$ over $\mathbf{Q}_0$ is p (the order of the factor group G/H) and the degree of $\mathbf{K}$ over $\mathbf{L}$ is h (the order of H). This is a direct generalization of Gauss' theory of the cyclotomic equation.

Suppose we can find in G a nested sequence of normal subgroups

$$G \supset H_1 \supset H_2 \supset \cdots \supset H_S \supset E \tag{10}$$

such that the factor groups $G/H_1, H_1/H_2, \ldots, H_S$ have prime orders p_1, $p_2, \ldots, p_S$ respectively. This turns out to be a necessary and sufficient condition for $\mathbf{K}$ to be obtainable from $\mathbf{Q}_0$ by a finite number of extensions

$$\mathbf{Q}_0 \subset \mathbf{L}_1 \subset \mathbf{L}_2 \subset \cdots \subset \mathbf{L}_S = \mathbf{K}.$$

Furthermore, all these extensions are radical extensions: $\mathbf{L}_1$ is obtained from $\mathbf{Q}_0$ by adjunction of a root of the equation $x^{p_1} - a = 0$, $a \in \mathbf{Q}_0$, $\mathbf{L}_2$ is obtained from $\mathbf{L}_1$ by adjunction of a root of the equation $x^{p_2} - a_1 = 0$, $a_1 \in \mathbf{L}_1$, and so on. Each of these equations has a cyclic Galois group of prime order and is therefore solvable by radicals (we are assuming that the necessary roots of unity have already been adjoined to $\mathbf{Q}_0$).

A group G is called solvable if it has a chain of normal subgroups (10) with the specified property. Thus a necessary and sufficient condition for the solvability of an equation by radicals is the solvability of its Galois group. In this way the investigation of the structure of an infinite field $\mathbf{K}$ has been reduced to the investigation of the structure of a finite group G.

Galois concluded with the investigation of equations of prime degree. For such equations he proved the following result: For an irreducible equation of prime degree p to be solvable by radicals it is *necessary and sufficient* that its group consist of permutations of the form

$$x_k \longrightarrow x_{ak+b},$$

where a takes on the values $1, 2, \ldots, p-1$ and b takes on the values $1, 2, \ldots, p$, i.e., the group has order $p(p-1)$ (of course, the values of $ak+b$ are taken mod p). This is the one-dimensional affine linear group modulo p considered earlier.[4]

We see that in order to obtain a criterion for the solvability of equations by radicals Galois constructed a complex chain of interrelated concepts (an "algorithm of concepts"): from the given equation it was necessary to go over to a normal equation by constructing a primitive element; to make precise the concept of a group of permutations; to define the Galois group of the equation; to introduce the concepts of a normal subgroup, of a factor group (Galois had no name for the last object), and of a solvable group. The fundamental question was settled not by transformations and computations but by the complex conceptual algorithm just described.

It is remarkable that in Galois theory groups turn up in the same context in which they turned up later in other theories. The context we have in mind is that of the set of invariants of a group. In the case of Galois theory the set of invariants of a group forms a field. The idea of a group and its invariants permeates all of 19th-century mathematics. Klein's classification of geometries was based on this idea and so was the theory of modular and automorphic functions (developed by Klein, Poincaré, Fuchs, and others).

4. The evolution of group theory in the 19th century

Galois' letter to Chevalier was published in September 1832 but went unnoticed. It was only in 1846, i.e., 14 years after his death, that the noted French mathematician Liouville collected all of Galois' papers, including his

Memoir on the Solvability of Equations by Radicals, provided them with commentaries, and published them in his *Journal des mathématiques pures et appliquées*. This was the beginning of a new phase in the life of Galois theory. As early as 1856, a complete account of the theory was presented in the third edition of Serret's *Course of Higher Algebra* (the account in the second edition of 1854 was incomplete), i.e., it became part of the textbook literature. By that time, German and English mathematicians, such as Kummer, Kronecker, and Cayley, were familiar with the theory.

When Galois' papers were first published many theorems of group theory were already known. We already mentioned Lagrange's theorem. We also note the group-theoretic character of many of Euler's proofs in number theory, such as Fermat's little theorem and its generalizations.

In his *Arithmetical Investigations* Gauss introduced an operation of composition of binary quadratic forms and thereby extended composition to objects very different from numbers.

We recall that a binary quadratic form is an expression of the form

$$f(x, y) = ax^2 + 2bxy + cy^2, \tag{11}$$

where $a, b, c \in \mathbf{Z}$. Gauss called the number $D = b^2 - ac$ the determinant of the form. The fundamental question considered by Fermat, Euler, and Lagrange was the determination of the range of the form (11), i.e., of the set M of integers N with the property that for every $N \in M$ there are integers x_0, y_0 such that

$$N = ax_0^2 + 2bx_0y_0 + cy_0^2.$$

Some early relevant results were obtained by Fermat. Thus Fermat claimed that the range of $x^2 + y^2$ includes all primes of the form $4n + 1$ and no integer of the form $4n + 3$, and that the range of $x^2 + 2y^2$ includes all primes of the form $8n + 1$ and $8n + 3$ and none of the form $8n + 5$ and $8n + 7$. He also obtained analogous criteria for the representability of primes by the forms $x^2 - 2y^2$ and $x^2 + 3y^2$.

Lagrange noted that if N is representable by a form (11) then it is also representable by the form f', obtained from (11) by a substitution invertible over the integers, i.e., a substitution

$$\begin{aligned} x &= \alpha x' + \beta y', \\ y &= \gamma x' + \delta y', \end{aligned} \tag{12}$$

where $\alpha, \beta, \gamma, \delta$ are integers and $\alpha\delta - \beta\gamma = \pm 1$. In that case, there exists a substitution, inverse to (12), that transforms f' into f. Gauss called forms f

and g related by a unimodular substitution (12) (i.e., one with $\alpha\delta - \beta\gamma = 1$) strictly equivalent, $f \sim g$.

As Gauss showed, the determinant D is invariant under a unimodular transformation. Forms with a given determinant belong to finitely many classes, a result first shown by Lagrange.

Gauss introduced a composition that associated with two given forms f_1 and f_2 with given determinant a third form f_3 with the same determinant

$$f_3 = f_1 \oplus f_2 \tag{13}$$

and extended this operation to classes of forms. In other words, he showed that if $f_1 \sim g_1, f_2 \sim g_2$, and $g_3 = g_1 \oplus g_2$, then $g_3 \sim f_3$. This meant that one could speak of composition of classes of strictly equivalent forms:

$$K \oplus L = M.$$

This was the first example of extension of operations defined for a set of objects to a quotient set. Gauss regarded a class of forms as a single object. He called the class E containing the form $x^2 - Dy^2$ principal and showed that it played the role of a unit:

$$K \oplus E = E \oplus K = K,$$

where K is an arbitrary class. He also showed that the composition of classes is associative and commutative and that the equation

$$K \oplus X = L$$

has a unique solution.[5]

Gauss remarked that the operation of composition of classes could be denoted by the $+$ symbol, i.e., he regarded it as analogous to an arithmetical operation. It would have been more natural to regard it as an analogue of multiplication rather than addition. According to Dieudonné, Gauss regarded the choice of a symbol as insignificant, and what he viewed as important was the idea of a law of composition.

Gauss' proofs are completely general. At no point does he make use of a special property of a class of forms. This being so, his argument could be applied without modification to a law of composition of objects of an arbitrary nature. He notes that the group of classes need not be cyclic: "we observe that, since in this case [the case discussed earlier] one base is insufficient, we must take two or more classes and from their multiplication and composition produce all the rest" [Art. 306, IX of *Disquisitiones*]. From a subsequent part of the *Disquisitiones* it becomes clear that (in modern terms) Gauss claims that the commutative group of classes of forms is a direct sum of two (or more) cyclic groups.[6]

Another research stream was connected with the study of groups of permutations. Already Lagrange showed that the order of a subgroup H of the symmetric group S_n divides $n!$. In 1815 there appeared Cauchy's "Memoir sur le nombres de valeurs qu'une fonction peut acquérir lorsqu'on y permute de toutes les manières possible les quantités qu'elle renferme" (*J. Ec. Polyt.* 1815) in which he investigated the number of values taken on by a rational function $f(x_1, x_2, \ldots, x_n)$ under all permutations of its arguments. To answer this question he embarked on a systematic construction of a theory of groups of permutations. This was the first time such groups became an independent object of study. He wrote a permutation as $\begin{pmatrix} a & b & c & \cdots & l \\ \alpha & \beta & \gamma & \cdots & \lambda \end{pmatrix}$, or briefly as $\begin{pmatrix} A \\ B \end{pmatrix}$, and introduced a composition of permutations

$$\begin{pmatrix} A \\ B \end{pmatrix} \begin{pmatrix} B \\ C \end{pmatrix} = \begin{pmatrix} A \\ C \end{pmatrix},$$

as well as the identity permutation $\begin{pmatrix} A \\ A \end{pmatrix}$. This paper marked the beginning of the evolution of the theory of groups (a term not used by Cauchy) of permutations.

Cauchy's papers contain the first general theorems on groups of permutations. In particular, he proved that if a nonsymmetric function of n quantities takes on fewer different values than the largest prime not exceeding n then it takes on exactly two different values. In 1845 Bertrand showed that for $n > 4$ the group S_n has no subgroup of index > 2 and $< n$.

Between 1844 and 1846 Cauchy returned to the study of groups of permutations and proved the following deep result: If the order of a group of permutations is divisible by a prime p then the group contains a subgroup of order p.

In 1872 Sylow (1832–1918) generalized Cauchy's result by showing that if the order of a group is divisible by p^k then the group contains a subgroup of order p^k.

As mentioned earlier, a turning point in the evolution of group theory was the publication of Galois' papers (1846). Thus Camille Jordan (1838–1922) began his investigations by providing commentaries on Galois' frequently very fragmentary papers. His famous *Traité des substitutions et des équations algébriques* (Paris, 1870) was, in a sense, a summary. It contained the first complete and systematic account of the theory of groups of substitutions as well as applications to geometry, the theory of elliptic functions, and algebra.

Jordan did not give a general definition of a group. He wrote: "We will say that a system of permutations forms a group or a sheaf (faisceau) if the product of any two permutations in the system belongs to the system", i.e., he defined a semigroup. But using the example of a group of permutations he essentially developed the general theory of finite groups. He explicitly introduced the notions of a normal subgroup, (which he called a "singular subgroup"), of a simple group, and of a homomorphism (a "meriedric isomorphism") and isomorphism (a "holoedric isomorphism") of groups. Jordan showed that the cosets with respect to a normal subgroup form a group, which he called in his *Treatise* a group "modulo a singular subgroup" and later (in 1873) a factor group. Finally, he introduced the notion of a composition series

$$G \supset G_1 \supset \cdots \supset G_m \supset e,$$

where G_{i+1} is normal in G_i and the factor groups G_i/G_{i+1} are of prime order. The concept of a composition series is of great importance in Galois theory. Jordan showed that two composition series of the same group always have the same number of terms and that (apart from order) the groups G_i/G_{i+1} in the two series have the same number of elements. In 1889 Hölder showed that (apart from order) the groups G_i/G_{i+1} in the two series are isomorphic.

It was Jordan who called commutative groups Abelian.

In a special chapter of his *Treatise* Jordan began to consider the representation of groups of permutations by matrices, more specifically, by linear subgroups of invertible square $n \times n$ matrices with elements in the fields $\mathbf{F}_q, \mathbf{R}$, or $\mathbf{C}$. The evolution of this theory took place at the end of the 19th and in the 20th centuries.

Already in 1854, also stimulated by Galois' papers, Arthur Cayley (1821–1895) arrived at the thought that the nature of the group elements is irrelevant to the study of the properties of groups. He was the first to introduce the notion of an abstract group. He did this in the two-part paper "On the Theory of Groups, as Depending on the Symbolic Equation $\Theta^n = 1$" (*The Collected Mathematical Papers*, 1854, Vol. 2, pp. 123–132). He did not require the group operation to be commutative but assumed that it was associative

$$(\alpha\beta)\gamma = \alpha(\beta\gamma).$$

He also stipulated that multiplication of all group elements by a group element must always yield all group elements. This implies the unique solvability of the equations $\alpha x = \beta$ and $y\alpha = \beta$ for each pair of group elements α, β.

Cayley clarified his definition by means of the group multiplication table

	1	α	β	γ	$\cdots$
1	1	α	β	γ	$\cdots$
α	α	α^2	$\beta\alpha$	$\gamma\alpha$	$\cdots$
β	β	$\alpha\beta$	β^2	$\gamma\beta$	$\cdots$
γ	γ	$\alpha\gamma$	$\beta\gamma$	γ^2	$\cdots$
$\cdots$	$\cdots$	$\cdots$	$\cdots$	$\cdots$	$\cdots$

Here each row and column contains all group elements $1, \alpha, \beta, \ldots$

Cayley introduced the term group in honor of Galois.

The groups given by Cayley as examples in his early papers were always finite. He prescribed them by multiplication tables and by generators and defining relations. He showed that there are two different (we would say nonisomorphic) groups of order 4, two of order 6, and 5 of order 8. He also showed that the only group of prime order p is the cyclic group

$$\{1, \alpha, \alpha^2, \ldots, \alpha^{p-1}\}.$$

Cayley's abstract formulation of the group concept was a reflection of the influence of the English school of "symbolical algebra" that came into being in the 1830s. It included such mathematicians as Boole and Hamilton. We will discuss the work of this school in Chapter IX.

Cayley's ideas did not meet with an enthusiastic reception. We saw that even in 1870 Jordan invariably operated with groups of permutations. What seems to have helped Cayley's approach was his paper on groups published in 1878. A few years later (in 1882) there appeared Walter von Dyck's paper "Investigations on the Theory of Groups", in which he introduced the notion of a free group, whose elements are "words" $A^r B^k C^l \ldots$ in a number of symbols $A, B, C, \ldots$.

The generally accepted axiomatics of group theory were introduced by Heinrich Weber in the first volume (1894) of his three-volume *Course of Algebra*. For a long time this work served as the standard reference for all mathematicians interested in algebra.

5. The victorious "march" of group theory

We have seen how the group concept, one of the most important concepts of modern mathematics, emerged gradually from investigations in algebra and number theory. The extraordinary progress of algebra, analysis, geometry, mechanics, and theoretical physics is due to the idea of a group and its associated set of invariants. In what follows, we note some of the stages of the victorious "march" of group theory. Needless to say, we limit ourselves to listing facts.

1. In 1870 Jordan classified all finite rotation groups in three-dimensional space. He constructed for linear differential equations a theory analogous to Galois theory, in which the role of the Galois group of an algebraic equation is played by the monodromy group of the differential equation (1871 and later). This theory was completed by Émile Picard.

2. In his Erlanger Programm of 1872 Felix Klein applied groups to the classification of geometries.

3. In the 1880s Poincaré introduced into topology the notion of a fundamental group. Then Poincaré, P. S. Aleksandrov, and Kolmogorov introduced homology groups.

4. In the 1880s Poincaré introduced groups into analysis for the study of one of the most important classes of functions, namely automorphic functions. Similar investigations were carried out by F. Klein and Hermann A. Schwarz.

5. Lie and Klein began to develop the theory of continuous groups, whose role in the theory of partial differential equations is analogous to that of groups of permutations in Galois theory.

6. Already Jordan began to consider the theory of representations of groups by matrices. This idea turned out to be particularly fruitful for the further evolution of algebra. A general theory of group representations was constructed in the 1890s by Fedor E. Molin (better known in the West as Theodor Molien) and Georg Frobenius.

7. Harmonic analysis, i.e., differential and integral calculus on topological groups, was first developed by Hermann Weyl, but its principal results are due to L. S. Pontryagin.

Soon groups entered physics. Here the first step was due to the crystallographers E. S. Fedorov (1853–1919) and A. M. Schoenflies (1853–1928). In 1890 Fedorov used groups to solve the problem of classification of regular systems of points in space (crystals). Schoenflies solved the same problem independently of Fedorov at roughly the same time. It turned out that there are 17 Fedorov groups in the plane and 230 Fedorov groups in space. Their detailed classification would have been impossible without the use of group theory.

Group theory plays a crucial role in quantum physics. Wolfgang Pauli, one of the creators of quantum physics, wrote that the ideas of group theory belong to "the most powerful instruments of modern physics", and that, in his view, their fruitfulness is "far from having been exhausted" (W. Pauli, *Collected Scientific Papers*, ed. R. Kronig and V. F. Weisskopf, 2 vols., Interscience, 1964). The subsequent evolution of physics brought with it a splendid confirmation of Pauli's words.

Editor's notes

[1] In this table Gauss uses $[\lambda]$ to denote r^λ, and the entry $(2,1) \ldots [1], [18]$ means that [1] and [18] are roots of the quadratic equation $x^2 - (2,1)x + 1 = 0$.

[2] A strong case can be made that Abel was motivated by the lemniscate. In his 1829 Memoir cited in the text, Abel explains that his theory applies to the cyclotomic equation, and then goes on to say: "The same property pertains to a certain class of equations which I obtained by the theory of elliptic functions." This refers to the equations satisfied by division points of an elliptic curve, which in the case of a curve with complex multiplication give an Abelian extension. It was Abel who discovered this in his great paper "Recherches sur les fonctions elliptiques" (*Crelle*, 1827–1828). One of the high points of this paper is his treatment of the question of when one can divide the lemniscate into n equal arcs using ruler and compass. He shows that one can do this for the same n's as for the circle. Also, in his 1829 Memoir, he says: "After having developed this theory in general, I will apply it to circular and elliptic functions." So it is clear that elliptic functions played a crucial role in Abel's 1829 Memoir.

[3] In 1853, Kronecker gave the name "Abelian equation" to the types of equations Abel introduced here, and then, in 1870, Jordan used the term "Abelian group" for the Galois group of an Abelian equation. Thus we use the term "Abelian" for commutative groups because of Galois theory and (see the previous comment) elliptic functions!

[4] To understand the structure of this group, note that the permutations $x_k \to x_{k+b}$ form a cyclic normal subgroup, and the quotient is cyclic in view of the existence of a primitive root. In general, a group is called *metacyclic* if it has a cyclic normal subgroup with a cyclic quotient, and every metacyclic group is clearly solvable. Thus the one-dimensional affine linear group modulo p is metacyclic and hence solvable.

[5] The last four paragraphs call for comments.
Ad equivalence: According to Gauss, two forms related by (12) are "equivalent" when $\alpha\delta - \beta\gamma = \pm 1$ and (as already noted) "strictly equivalent" when

it is 1. He called the classes of forms associated with these two cases "equivalence classes" and "strict equivalence classes". This example gave rise to the general notion of equivalence class.

Ad composition: Gauss defined both "composition" and "direct composition". The former is a generalization of formulas such as (15) on p. 9 and $(*)$ on p. 33. The latter is a refinement of composition which is needed for the group structure on strict equivalence classes. Thus equation (13) uses direct composition. In contrast, Legendre considered the binary operation on equivalence classes given by composition. This led to a multivalued group operation since he was not using direct composition and strict equivalence.

[6] Ad "multiplied and composed": Given classes $A, B, C, \ldots$, the group operation is written additively, so that these classes generate $nA + mB + kC + \cdots$, where $n, m, k, \ldots$ are integers. Thus one "multiplies" to get $nA, mB, kC, \ldots$, and then "composes" to get $nA + mB + kC + \cdots$.

CHAPTER 8

Problems of number theory
and the birth of commutative algebra

1. Diophantine equations and the introduction of algebraic numbers

We saw that many of the investigations in algebra were connected with the problem of solution of equations by radicals. These investigations stimulated the creation and evolution of the theories of groups and fields. Other algebraic investigations are rooted in Diophantine analysis and in number theory; more specifically, in Fermat's last theorem and in the reciprocity laws.

Euler proved Fermat's last theorem for $n = 4$ in 1738 but it took another 30 years for him to prove it for $n = 3$. His letters to Goldbach show that the proof cost him a great deal of effort because it called for the introduction of new methods.

As early as the 1760s Euler and Lagrange began to apply irrational as well as imaginary expressions of the form $a + b\sqrt{c}$, $a, b, c \in \mathbf{Z}$, to problems in number theory. In this connection Euler wrote to Lagrange: "I was delighted by your method of using irrational as well as imaginary numbers in the part of analysis devoted to rational numbers alone. Similar ideas occurred to me a few years ago ... in connection with the publication of a complete algebra in Russian, I presented this method in detail and showed that to solve the equation

$$x^2 + ny^2 = (p^2 + nq^2)^\lambda$$

it suffices to solve the equation

$$x + y\sqrt{-n} = (p + q\sqrt{-n})^\lambda \text{ "}$$

(Lagrange, *Oeuvres*, Vol. 14, p. 215).

It seems that at that time it occurred to Euler to use expressions of the form $a + b\sqrt{c}$ to solve Fermat's last theorem. However, to do this one had to take another decisive step, namely, to regard these expressions as integers.

What does this mean? In what way do the integers differ from, say, the rational numbers? Primarily by having a rich arithmetic. For example, there are prime and composite integers, and every composite integer can be written in an essentially unique way as a product of primes. (When it comes to the natural numbers, this law was known (in a somewhat restricted form) as far back as Euclid.) One consequence of this law is that if the product of two relatively prime positive integers is equal to a certain power of an integer then so are those integers: if $(a, b) = 1$, $a, b > 0$, and $ab = l^k$ then $a = t_1^k$ and $b = t_2^k$. Euler carried over all these properties, without proof, to numbers of the form $p + q\sqrt{-3}$, $p, q \in \mathbf{Z}$, and used them to prove Fermat's last theorem for $n = 3$. We give his proof.

Suppose that there are three integers x, y, z such that

$$x^3 + y^3 = z^3.$$

We can assume that they are pairwise relatively prime. Then two of them must be odd and the third even (otherwise they would have a nontrivial common factor). Suppose that x and y are odd and z is even (otherwise we would prove the impossibility of $x^3 = y^3 - z^3$). Putting $x = p + q$ and $y = p - q$, where p and q have different parities and $(p, q) = 1$, we have

$$x^3 + y^3 = 2p(p^2 + 3q^2) = z^3.$$

Since z^3 is even, it is divisible by at least 8. Also, p is even, q is odd, and

$$\frac{p}{4}(p^2 + 3q^2) = z_1^3.$$

Euler considered two cases: 1) p is not divisible by 3; 2) p is divisible by 3. To understand Euler's guiding idea it suffices to consider the first of these two cases. If p is not divisible by 3, then $p/4$ and $p^2 + 3q^2$ are relatively prime and thus are both cubes. Now—and this is the crucial step in the proof—Euler factors $p^2 + 3q^2$:

$$p^2 + 3q^2 = (p + q\sqrt{-3})(p - q\sqrt{-3}) = r^3.$$

From this he concludes that each of the imaginary factors is a cube,

$$p \pm q\sqrt{-3} = (u \pm v\sqrt{-3})^3.$$

Hence
$$p = u(u - 3v)(u + 3v);$$
$$q = 3v(u + v)(u - v).$$

Since q is odd, so is v. Hence u is even. Since $p/4$ is a cube, so is $2p$, i.e., $2u(u - 3v)(u + 3v) = t^3$. Since $2u, u - 3v$, and $u + 3v$ are relatively prime, $2u = t_1^3$,

$$u - 3v = f^3$$
$$\underline{u + 3v = g^3}$$
$$2u = f^3 + g^3 = t_1^3.$$

It is easy to see that f, g, t_1 are smaller than x, y, z respectively. This implies that if our initial equation had a (positive) integral solution x, y, z then it would have a smaller solution f, g, t_1. But then we could find a solution f_1, g_1, t_2 smaller than the solution f, g, t_1, and so on. This is impossible, for there are only finitely many whole numbers smaller than a given one. This is an application of Fermat's method of descent. What is new in this argument is that Euler carried over the divisibility laws from the whole numbers to expressions of the form $p + q\sqrt{-3}$.

We note that Euler's reasoning was not rigorous. Furthermore, it involved a false assumption. Indeed, the example

$$4 = 2 \cdot 2 = (1 + \sqrt{-3})(1 - \sqrt{-3})$$

shows that the ring of integers of the form $m + n\sqrt{-3}$, $m, n \in \mathbf{Z}$, is not a unique factorization domain. Unique factorization holds in the maximal ring of integers of the field $Q(\sqrt{-3})$ whose elements are of the form $(m + n\sqrt{-3})/2$, $m \equiv n \pmod 2$. The reason Euler managed to avoid mistakes was that he worked with numbers of the form $p + q\sqrt{-3}$, $p \not\equiv q \pmod 2$.

Euler's proof contained two important ideas that were subsequently adopted by mathematicians. The first of these ideas was that when proving Fermat's last theorem one must write the form $x^\lambda + y^\lambda$ (λ prime) as a product of linear factors:

$$x^\lambda + y^\lambda = (x + y)(x + \zeta y) \cdots (x + \zeta^{\lambda - 1} y),$$

where $\zeta^\lambda = 1$, $\zeta \neq 1$.

The second, more important, idea was that in order to investigate the properties of ordinary integers it was necessary to extend the notion of an integer. In the 19th century the development of this idea gave rise to the creation of the theory of algebraic numbers and to the construction of their arithmetic.

Euler did not justify the application of rules of ordinary arithmetic to expressions of the form $m + n\sqrt{-3}$, $m, n \in \mathbf{Z}$.

The first rigorous introduction of algebraic integers is due to Gauss. He did this between 1828 and 1832 in his famous paper "The Theory of Biquadratic Residues" (Gauss, *Untersuchungen über höhere Arithmetik*, Berlin, 1889. Repr. by Chelsea, 1965, pp. 511-586). Gauss realized that extension of the law of quadratic reciprocity (for which he gave eight proofs, at a time when none existed) to a law of biquadratic reciprocity required an extension of the very notion of an integer. For more than 2000 years this notion seemed to be inseparable from the domain $\mathbf{Z}$ of rational integers; it seemed to be its intrinsic property. Gauss separated the notion of an integer from its natural carrier and transferred it to the ring of numbers of the form

$$a + bi, \tag{1}$$

where $a, b \in \mathbf{Z}$ and i is a root of the equation

$$x^2 + 1 = 0, \tag{2}$$

which is irreducible over $\mathbf{Q}$. He showed that the domain $\mathbf{O}$ of the numbers in (1), now known as Gaussian integers, is closed under addition, subtraction, and multiplication, and that it is possible to set up in it a structure analogous to ordinary arithmetic. He defined for the new numbers the notion of a unit (there are four of them, namely, $1, -1, i, -i$), called numbers obtainable from one another by multiplication by a unit associates, and stated that one should not distinguish factorizations which differ by associates. He called a number in $\mathbf{O}$ a prime if it could not be written as a product of two factors neither of which is a unit, and called the rational integer $a^2 + b^2 = (a + bi)(a - bi)$ the norm $N\alpha$ of $\alpha = a + bi$ in (1). It follows from the definition that

$$N\alpha\beta = N\alpha N\beta.$$

It is clear that a composite rational integer is a composite Gaussian integer but that a prime rational integer may be a composite Gaussian integer. Thus $2 = (1 + i)(1 - i)$ and $5 = (2 + i)(2 - i)$. More generally, every rational prime of the form $4n + 1$ (such primes can be written as sums of squares) is a composite Gaussian integer:

$$p = m^2 + n^2 = (m + ni)(m - ni).$$

On the other hand, as Gauss shows, primes in $\mathbf{Z}^+$ of the form $4n + 3$ remain prime in $\mathbf{O}$ and the norm of a number q of this kind is q^2.

Gauss showed that a Gaussian integer $a + bi$, $ab \neq 0$, is prime or composite according as its norm is a prime or composite rational integer. This implies that the Gaussian primes are the associates of the following: 1) the (associated) divisors $1 + i$ and $1 - i$ of 2; 2) the rational primes of the form $4n + 3$; and 3) the (conjugate) divisors $a + bi$ and $a - bi$, $b \neq 0$, of the rational primes of the form $4n + 1$.

Next Gauss proved that the Gaussian integers were a unique factorization domain.

After this, he easily developed in **O** an arithmetic analogous to that in **Z** (he proved Fermat's ("little") theorem, introduced congruences, primitive roots, and so on) and applied it to state, and partially prove, the law of biquadratic reciprocity. In this way he showed the mathematical world that complex integers were just as "respectable" objects of higher arithmetic as rational integers.

The impact of Gauss' paper was tremendous. One indication of this impact was that up to the 1860s algebraic integers were referred to as complex integers even if they were of the form $a + b\sqrt{D}$, $D > 0$, i.e., even if they were real.

We note that Gauss' paper played a decisive role in getting mathematicians to regard expressions of the form $a + bi$ as numbers. Indeed, it became clear that they could be objects of (an extended) arithmetic, and that, moreover, this extended arithmetic could be used to obtain results about rational integers that could not be obtained without it.

Gauss was well aware that his paper opened before mathematicians a boundless world of problems. He wrote that "In some sense of the word, this theory (i.e., the law of quadratic reciprocity (*the authors*)) is bound to extend infinitely the domain of higher arithmetic" (Gauss, ibidem, p. 511), and that "The natural source of the general theory must be sought in the extension of the domain of arithmetic" (Gauss, ibidem, p. 540).

He noted that the theory of cubic residues (and the law of cubic reciprocity) must be based on the investigation of numbers of the form $a + b\rho$, where $\rho^3 - 1 = 0$ and $\rho \neq 1$ and added that "The theory of residues of higher degree requires the introduction of other imaginary quantities" (Gauss, ibidem, p. 541).

In the 1840s Eisenstein, Dirichlet, and Hermite independently defined algebraic integers as roots of algebraic equations of the form

$$F(x) = x^n + a_1 x^{n-1} + \cdots + a_n = 0, \quad a_1, \ldots, a_n \in \mathbf{Z}. \qquad (3)$$

Eisenstein showed that the sum and product of algebraic integers were again algebraic integers. Finally, in 1848, Dirichlet was able to determine the group of units in an arbitrary field of algebraic numbers. (*Note.* A field of algebraic numbers is a set $\mathbf{K} = \mathbf{Q}(\theta)$ of expressions of the form

$$\alpha = b_0 + b_1\theta + \cdots + b_{n-1}\theta^{n-1}, \tag{$*$}$$

where θ is a root of an irreducible equation of degree n and the $b_i \in \mathbf{Q}$. It is easy to see that the sum, product, and quotient (with nonzero denominator) of numbers of the form $(*)$ are again numbers of the form $(*)$.)

In 1847 a remarkable event occurred in the Paris Academy of Science. During the meeting that took place on 1 March, the French mathematician Lamé presented a memoir ostensibly containing a complete proof of Fermat's last theorem. Like his predecessors (Euler for $n = 3$, Lagrange, Gauss, and Legendre for $n = 5, 7$), Lamé wrote the equation $x^\lambda = z^\lambda - y^\lambda$ as the product

$$z^\lambda - y^\lambda = (z - y)(z - \zeta y) \cdots (z - \zeta^{\lambda-1}y), \tag{4}$$

where λ is a prime (it is clear that it is enough to prove Fermat's last theorem for prime exponents), $\zeta^\lambda = 1$, and $\zeta \neq 1$. The proof involved operating with numbers of the form

$$b_0 + b_1\zeta + \cdots + b_{\lambda-2}\zeta^{\lambda-2}, \quad b_i \in \mathbf{Z}, \tag{5}$$

under the assumption that these numbers obey the laws of ordinary arithmetic, including the law of unique factorization.

This assumption attracted the attention of Liouville who made the following comment: "The idea of introducing C. N. (i.e., complex numbers (*the authors*)) into the theory of the equations $x^n - y^n = z^n$ is not new; quite naturally, it is bound to occur to geometers owing to the binary form $x^n - y^n$. But I have not been able to deduce from this a satisfactory proof. At any rate, the attempts I made showed me that, as a first step, it was necessary to establish for C. N. a theorem analogous to the elementary theorem for whole numbers which asserts that a product can be decomposed into prime factors in just one way. Lamé's analysis only confirms me in this conviction. Is there no gap here that must be filled?" (quoted after R. Nogues, *Fermat's Theorem*. Paris, 1966, p. 38). As a result of Liouville's remark, the arithmetic of fields of algebraic numbers was, for an admittedly short time, at the very center of attention of French mathematicians. It was studied by Lamé, Wantzel, and even Cauchy. During the meeting of the Academy that took place on 22 March 1847 Cauchy presented a memoir in which he tried to deduce the Euclidean algorithm for numbers of the form (5). The memoir was printed in installments in the *Comptes rendus* for 1847. In the last installment Cauchy arrived at the conclusion that this could not be done.

2. Kummer's ideal factors

A few years earlier, numbers of the form (5) were studied by the eminent German mathematician Ernst Kummer (1810–1893). Kummer came across these numbers when he tried to generalize and prove the law of reciprocity for residues of arbitrary prime degree and to prove Fermat's last theorem, which he studied from 1837 on.

In 1846 Kummer noticed the amazing fact that in the ring $\mathfrak{O}$ of integers of the field $\mathbf{Q}(\zeta)$ it is possible for a product of two irreducible numbers α and β of the form (5) to be divisible, in a nontrivial way, by a third irreducible number γ of this form.

We explain the sense of this discovery in greater detail. It is well known that a prime p in the ring $\mathbf{Z}$ has the following two properties:

A. p cannot be written as a product of two factors none of which is a unit.

B. If a product ab is divisible by p, then at least one of the factors a and b is divisible by p.

Usually one takes the first of these properties as the definition of a prime and proves the second. But one can reverse this order, i.e., in the ring $\mathbf{Z}$ properties A and B are equivalent. Kummmer discovered that this is false for numbers of the form (5): they can have the first of these properties without the second. This put in doubt the possibility of constructing an arithmetic of the integers of the field $\mathbf{Q}(\zeta)$.

We give an example of nonunique factorization of algebraic integers.[1] In fields $\mathbf{Q}(\zeta)$ the nonuniqueness first shows up for $\lambda = 23$. This being so, we will take an example from the quadratic field $\mathbf{Q}(\sqrt{-5})$. Here

$$6 = 2 \cdot 3 = (1 + \sqrt{-5})(1 - \sqrt{-5})$$

and none of the numbers $2, 3, 1 + \sqrt{-5}$, and $1 - \sqrt{-5}$ is representable as a (nontrivial) product of two factors of the form $a + b\sqrt{-5}$.

Kummer managed to save the situation, i.e., he made possible the construction of an arithmetic of the integers of the fields $\mathbf{Q}(\zeta)$ by the introduction of new objects, which he called ideal factors. In the spring of 1847 he wrote about his discovery to Liouville and presented the complete theory in two papers: "On the Theory of Complex Numbers" ("Zur Theorie der complexen Zahlen," J. für Math., 1847) and "On the Decomposition of Complex Numbers

Formed from Roots of Unity into Prime Factors" ("Über die Zerlegung der aus Wurzeln der Einheit gebildeten complexen Zahlen in ihre Primfactoren", ibidem).

Kummer's idea was that integers of the field $\mathbf{Q}(\zeta)$ that have property A but not property B are not genuine primes; they are products of factors not found in $\mathbf{Q}(\zeta)$ in "pure form". Thus in our example

$$2 = \mathfrak{p}_1\mathfrak{p}_2, \quad 3 = \mathfrak{q}_1\mathfrak{q}_2, \quad 1 + \sqrt{-5} = \mathfrak{p}_1\mathfrak{q}_1, \quad 1 + \sqrt{-5} = \mathfrak{p}_2\mathfrak{q}_2.$$

Here we could put

$$\mathfrak{p}_1 = 1 + i, \quad \mathfrak{p}_2 = 1 - i, \quad \mathfrak{q}_1 = \frac{1 + \sqrt{-5}}{1 + i} \quad \mathfrak{q}_2 = \frac{1 - \sqrt{-5}}{1 - i}.$$

All of these factors are algebraic integers (true, they do not belong to the field $\mathbf{Q}(\sqrt{-5})$; this is obvious for $\mathfrak{p}_1$ and $\mathfrak{p}_2$ and easy to show for $\mathfrak{q}_1$ and $\mathfrak{q}_2$ by constructing equations of the form (3) with rational coefficients which they satisfy).

Kummer's methods for the introduction of ideal factors were later called local because they yielded ideal factors not for the whole ring of integers of the field $\mathbf{Q}(\zeta)$ at once but individually, for one prime p at a time. We cannot give here a detailed description of these methods and so limit ourselves to the remark that they were based on the parallelism between the factorization of the polynomial

$$\Phi(x) = \frac{x^\lambda - 1}{x - 1} = x^{\lambda-1} + x^{\lambda-2} + \cdots + 1$$

mod p and the factorization of the number p itself into (real or ideal) prime factors in $\mathbf{Z}(\zeta)$.

We illustrate our remark by considering the simplest case in which p is of the form $m\lambda + 1$. Kummer showed that in this case $\Phi(x)$ is factorable mod p into linear factors,

$$\Phi(x) \equiv \prod_{k=1}^{\lambda-1} (x - u_k) \pmod{p},$$

and p is either factorable into $\lambda - 1$ primes in $\mathbf{Z}[\zeta]$ or else we associate with it $\lambda - 1$ ideal factors:

$$p = p_1 p_2 \cdots p_{\lambda-1}.$$

Also, $\alpha(\zeta) = b_0 + b_1\zeta + \cdots + b_{\lambda-2}\zeta^{\lambda-2}$ is divisible by p_i if

$$\alpha(u_i) \equiv 0 \pmod{p}.$$

The construction of ideal factors for other types of primes is somewhat more complicated.

The development of local methods was advanced by E. I. Zolotarev (1847–1878) and by Kurt Hensel (1861–1941), a member of the Kummer school who introduced the p-adic numbers.

Kummer showed that the introduction of ideal factors into the ring $\mathbf{Q}[\zeta]$ restored uniqueness of factorization of its integers and constructed for these "supplemented" rings arithmetics analogous to the usual arithmetic. He also proved Fermat's last theorem for all prime exponents λ that do not enter into the numerators of the first $(\lambda - 3)/2$ Bernoulli numbers, and formulated and proved the reciprocity law for power residues with prime exponents (he did the latter in papers which appeared during 1858–1887).

Fermat's last theorem was proved in 1995 by A. Wiles and R. Taylor who made use of its deep connections with algebraic geometry.

3. Arithmetic in arbitrary fields of algebraic numbers. Ideal theory

Kummer's papers showed the fruitfulness of constructing arithmetics of the integers in cyclotomic fields. This influenced mathematicians to try to construct an arithmetic of the integers in an arbitrary field $\mathbf{Q}(\theta)$, where θ is a root of an equation of the form (3).

This turned out to be a very difficult task. The first difficulty was that of providing a description of the integers of a field $\mathbf{Q}(\theta)$, where the term "integer" was interpreted in the sense of Eisenstein, Hermite, and Dirichlet as a root of an equation of the form (3) with integer coefficients. The integers of fields $\mathbf{Q}(\theta)$, encountered hitherto by Dirichlet, Eisenstein, Cauchy, and Kummer, were all of the form

$$b_0 + b_1\theta + \cdots + b_{n-1}\theta^{n-1}, \quad b_i \in \mathbf{Z}.$$

But it turned out that these fields contained other integers as well. For example, $(1+\sqrt{-3})/2$ is an integer because it satisfies the equation $x^2 - x + 1 = 0$. The first exhaustive solution of this problem was given by Richard Dedekind (1831–1916) in his famous Xth Supplement to the second edition of Dirichlet's *Lectures on Number Theory* (1871). (*Authors' note.* We wish to point out that already Newton essentially settled this question for the case of real quadratic fields in his *Universal Arithmetic*. Specifically, Newton showed that if

$$D \equiv 2 \text{ or } 3 \pmod 4,$$

then in $\mathbf{Q}(\sqrt{D})$ the integers are of the form

$$m + n\sqrt{D}, \quad m, n \in \mathbf{Z},$$

and if

$$D \equiv 1 \pmod 4,$$

then the integers are of the form

$$(m + n\sqrt{D})/2, \quad m \equiv n \pmod 2, \quad m, n \in \mathbf{Z}.$$

However, these investigations of Newton were overlooked in the 19th century.) Dirichlet showed that numbers of the form $b_0 + b_1\theta + \cdots + b_{n-1}\theta^{n-1}$ can be integers if the b_i are rational numbers such that the primes in their denominators divide the field discriminant. He showed further that it is always possible to construct a basis (later called minimal) $\omega_1, \omega_2, \ldots, \omega_n$, ω_i integers in $\mathbf{Q}(\theta)$, such that every integer of $\mathbf{Q}(\theta)$ is a linear combination of the ω_i with rational integral coefficients. Zolotarev also ran into this problem and gave a different, local, solution (1877).

Another, far more serious, difficulty encountered in the transition from cyclotomic fields to general fields of algebraic numbers was that the introduction of ideal factors, based on the parallelism between the factorization of the defining equation $F(x) = 0$ mod p and the factorization of p itself in the ring of integers of the field $\mathbf{Q}(\theta)$, could not be carried over to arbitrary fields of algebraic numbers without new and essential modifications. The difficulty involved was explained by Dedekind as well as by Zolotarev.

Zolotarev developed a general theory of divisibility by modifying Kummer's methods (1878; published in 1880). Another approach to this theory was proposed by Kummer's student Leopold Kronecker (1823–1891), whose memoir was published in 1882. Dedekind abandoned congruences altogether and solved the problem in a very different way. While the methods of Zolotarev and Kronecker turned out to be extremely effective in the study and solution of problems of number theory and algebraic geometry, Dedekind's method turned out to be of fundamental importance in general algebra and marked the beginning of a fundamental transformation of this branch of mathematics. That is why we will present Dedekind's construction of arithmetic in a field of algebraic numbers.

Dedekind's approach can be characterized as set-theoretic and axiomatic. His basic idea was to replace each of Kummer's ideal factors by the set $\mathbf{J}$ of integers of $\mathbf{Q}(\theta)$ that are divisible by this factor. The set $\mathbf{J}$ was defined in a manner independent of the notion of an ideal factor. Also, the number-theoretic notions (divisor, multiple, etc.) were replaced by set-theoretic notions (subset, inclusion, intersection, etc.). All this was done by Dedekind in the Xth Supplement to the second edition of Dirichlet's *Lectures on Number Theory*. This was the first work in which basic objects of algebra were introduced axiomatically. According to Bourbaki, the Supplement was written in a "general manner and in a completely new style."

Given the tremendous importance for algebra of Dedekind's new conception, we will try to describe the Xth Supplement in some detail. Dedekind begins by introducing the concept of a field (Körper):

"A field is a system of infinitely many real or complex numbers so closed and complete that the addition, subtraction, multiplication, or division of two of its numbers always yields a number of this system" (R. Dedekind, *Gesammelte mathematische Werke*. Braunschweig, 1930–1932, Vol. 3, p. 224).

We see that this definition differs from a modern one by designating the field elements as real or complex numbers. However, most of the consequences deduced by Dedekind are of a completely general character. He notes that the rationals are the smallest field, and all complex numbers, the largest. Then he introduces the notions of a subfield, of a basis, and of the degreee of a field.

Going over to the construction of arithmetic in the ring (Dedekind calls it order (Ordnung)) $\mathfrak{O}$ of integers of a field of algebraic numbers, Dedekind defines the notion of a module, destined to play a crucial role in algebraic number theory and in modern algebra. Finally he introduces the notion of an ideal, which he defines axiomatically: "A system $\mathfrak{A}$ of infinitely many numbers of $\mathfrak{O}$ is an ideal if it satisfies two conditions.

1. The sum and difference of any two numbers in $\mathfrak{A}$ is again a number in $\mathfrak{A}$.

2. The product of a number in $\mathfrak{A}$ by a number in $\mathfrak{O}$ is again a number in $\mathfrak{A}$" (ibidem, p. 251).[2]

We note that if γ is a nonzero number in $\mathfrak{O}$, then the set of all numbers $J(\gamma)$ in $\mathfrak{O}$ divisible by γ satisfies these two conditions and so is an ideal. Dedekind calls it a principal ideal. He defines divisibility of numbers and ideals in set-theoretic terms. Specifically, he says that a number α is divisible by an ideal $\mathfrak{A}$ if $\alpha \in \mathfrak{A}$, and that $\alpha \equiv \beta \pmod{\mathfrak{A}}$ if $\alpha - \beta \in \mathfrak{A}$.

If $\mathfrak{A}$ and $\mathfrak{B}$ are ideals and $\mathfrak{A} \subset \mathfrak{B}$, then Dedekind says that $\mathfrak{A}$ is divisible by $\mathfrak{B}$, or that $\mathfrak{B}$ is a divisor of $\mathfrak{A}$. He calls the intersection of two ideals their least common multiple, and the system of numbers of the form $\alpha + \beta$, $\alpha \in \mathfrak{A}$, $\beta \in \mathfrak{B}$, their greatest common divisor. Finally, he calls an ideal $\mathfrak{p}$ prime if it has no divisors other than $\mathfrak{O}$ and $\mathfrak{p}$.

Then Dedekind proves the fundamental theorem of divisibility theory: if $\alpha\beta \equiv 0 \pmod{\mathfrak{p}}$, where $\mathfrak{p}$ is prime, then at least one of the numbers α, β is divisible by $\mathfrak{p}$.

Briefly, the axiomatic method was introduced into algebra through Dedekind's papers. Initially, the new concepts were defined by this method only for one class of objects, namely, for fields of algebraic numbers.

4. Ideal theory in fields of algebraic functions

The introduction of such vital concepts as field, module, and ideal was just the first step on the road of formation of commutative algebra.

The second, equally important step, was the transfer of the whole theory to fields of algebraic functions. This was done by Dedekind and Heinrich Weber (1843–1913) in their joint work "The Theory of Algebraic Functions of a Single Variable" ("Theorie der algebraischen Functionen einer Veränderlichen", J. für Math., 1882).

In the 19th century, the investigation of fields of algebraic numbers, stimulated by the discovery of nonuniqueness of factorization in such fields, was paralleled by the development of the theory of algebraic functions. Here the key problem was dealing with multivalued functions of a complex variable. The problem was treated by Abel and Jacobi (1804–1851) but its definitive solution was due to Bernhard Riemann (1826–1866). To make a multivalued function singlevalued Riemann considered it on a multisheeted surface to each of whose points there corresponded a single value of the multivalued function. Such surfaces are now known as Riemann surfaces. But the construction of a Riemann surface involved continuity considerations based on geometric intuition. This could not satisfy Dedekind and Weber. In their memoir they said that their aim was "to justify the theory of algebraic functions of a single variable, which is one of the main achievements of Riemann's creative work, from a simple as well as rigorous and completely general viewpoint." Their construction is remarkable in that it is applicable to an arbitrary algebraically closed base field of characteristic 0 (i.e., they have no need of continuity!). They wrote: "For example, no gap would arise anywhere if one wished to restrict the domain of employed numbers to the system of algebraic numbers" (R. Dedekind, *Gesammelte mathematische Werke*, Vol. 1, p. 242).

In their theory Dedekind and Weber used the analogy between algebraic numbers and algebraic functions noted centuries earlier. As far back as the 16th century, Simon Stevin (1548–1620) observed that polynomials in one variable behave like integers, and irreducible polynomials like primes. He introduced the Euclidean algorithm for polynomials, which can be used to prove that they are a unique factorization domain. Gauss carried over to polynomials the theory of congruences. But it was the Dedekind–Weber memoir that demonstrated the full depth of the analogy between the two systems.

The memoir is divided into two parts. The first part is devoted to a formal theory of algebraic functions. Consider an irreducible equation

$$F(w, z) = a_0 w^n + a_1 w^{n-1} + \cdots + a_n = 0, \tag{6}$$

where the a_i are polynomials in z.

This equation determines a (usually multivalued) function w that Dedekind and Weber call an algebraic function. They use the "Kronecker construction" to form the field of algebraic functions Ω with elements

$$\zeta = b_0 + b_1 w + \cdots + b_{n-1} w^{n-1},$$

where the b_i are rational functions in z with coefficients in the base field. The degree of Ω over the field $\mathbf{C}(z)$ is n.

Then the authors carry over almost literally Dedekind's theory to fields of algebraic functions: they introduce the notion of an integral function of the field and study the ring of such functions, define a module (Funktionenmodul), introduce congruences with respect to a module, and, finally, define the notion of an ideal and prove the fundamental theorem of divisibility theory.[3] [Note [3] explains some of the terms used in the subsequent text. (Ed.)]

As an example, we state their definition of a module: "A system of functions (of Ω) is called a module if it is closed under addition, subtraction, and multiplication by every function integral with respect to z" (ibidem, pp. 251–252).

Of greatest interest is their definition of a point of the Riemann surface corresponding to the field of functions Ω. This issue is dealt with at the beginning of the second part of the memoir. The sequence of arguments involved is typical of Dedekind's approach. Suppose we had a point P of the surface in question. By evaluating the field functions at that point we would obtain a mapping of Ω into the field $\mathbf{C}$ of constants:

$$F \to F(P) = F_0 \in \mathbf{C}.$$

If $F \to F_0$ and $G \to G_0$, then it is clear that

$$F \pm G \to F_0 \pm G_0; \quad FG \to F_0 G_0; \quad F/G \to F_0/G_0.$$

The authors note that for the sake of generality it is reasonable to supplement $\mathbf{C}$ by ∞, for which one defines the usual operations of arithmetic except that no meaning is assigned to the symbols $\infty \pm \infty$, $0 \cdot \infty$, $\frac{0}{0}$, and $\frac{\infty}{\infty}$. Now consider all homomorphisms of Ω into the extended domain **C**. With each of these homomorphisms we can associate a point P. This motivated the definition employed by Dedekind and Weber: "Suppose that to all elements $\alpha, \beta, \gamma, \ldots$ of the field there correspond definite numerical values $\alpha_0, \beta_0, \gamma_0, \ldots$ such that (I) $\alpha_0 = \alpha$ if α is a constant and, in general, (II) $(\alpha + \beta)_0 = \alpha_0 + \beta_0$, (III) $(\alpha - \beta)_0 = \alpha_0 - \beta_0$, (IV) $(\alpha\beta)_0 = \alpha_0\beta_0$, (V) $(\alpha/\beta)_0 = \alpha_0/\beta_0$. Then we associate with the set of these values a point P.... We say: $\alpha = \alpha_0$ at P if α has at P the value α_0. Two points are said to be different if and only if there is a function in Ω that takes on different values at these points" (ibidem, p. 294).

The authors note that this definition is an invariant of the field Ω, for it is independent of the choice of independent variable used to represent a function of the field.

Before constructing a Riemann surface out of the points the authors prove the following theorems.

1. If $z \in \Omega$ is nonconstant and z takes on a finite value at P, then the functions in Ω integral with respect to z, that vanish at P, form a prime ideal in the ring of functions of Ω integral with respect to z.

The authors say that the point P generates a prime ideal $\mathfrak{p}$. If ω is a function integral with respect to z, then ω takes on at P a finite value ω_0 and they say that

$$\omega \equiv \omega_0 \pmod{\mathfrak{p}}.$$

2. Two different points cannot generate the same prime ideal.

3. If $z \in \Omega$ and $\mathfrak{p}$ is a [nonzero] prime ideal (relative to z), then there is one (and according to the preceding result) only one point P that generates this ideal; it is called a null point of the ideal $\mathfrak{p}$.[4]

These theorems imply the following construction of a Riemann surface T: Take any function $z \in \Omega$, form the ring of functions in Ω integral with respect to z, and consider all of its prime ideals $\mathfrak{p}$ and their corresponding null points P. In this way we obtain the points of the Riemann surface T_1 at which the function z is finite. To obtain the remaining points P_1 (at which $z = \infty$) take the function $z' = 1/z$, which vanishes at these points, form the ring of functions integral with respect to z' and construct its prime ideals containing z. By adding the new points $T_1{}'$ corresponding to the ideals $\mathfrak{p}'$ to the points T we obtain all points of the Riemann surface $T = T_1 \cup T_1{}'$.

The Dedekind–Weber construction has been extensively used in our own time in algebraic geometry, namely in the theory of schemes. Suppose that we are given an arbitrary commutative ring A with identity and we want to associate with it a natural geometric construct. In the theory of schemes we consider *all prime ideals* of A and call this set its *spectrum* (Spec(A)). [If A is an integral domain then Spec(A) includes the zero ideal.] Each of the prime ideals is called a *point of the spectrum*. For details see I. R. Shafarevich, *Basic Algebraic Geometry*, Springer, 1974.

It is well known that the axiomatic method in geometry was an achievement of antiquity and was the basis of Euclid's *Elements* of the third century BCE. The axiomatic method was introduced into algebra more than 2000 years later. We saw that this was connected with profound transformations of algebra, with the shift of its focus from the study of equations and elementary transformations to the study of algebraic structures defined on sets of objects of arbitrary nature.

We note that the context for Dedekind's study of "dual groups", later known as lattices, was his program of creation of abstract algebra. In the papers "On the Decomposition of Numbers through Their Greatest Common Divisors" (Über Zerlegungen von Zahlen durch ihre größten gemeinsamen Teiler". R. Dedekind, *Gesammelte mathematische Werke*, Vol. III, pp. 103–147) and "On Dual Groups Generated by Three Modules" (Über die von drei Moduln erzeugte Dualgruppen", ibidem, pp. 236–271), he defined a "dual group" for objects of arbitrary nature and studied its properties on the basis of explicitly formulated axioms. In the second of these papers Dedekind observes first that if we denote the greatest common divisor of modules $\mathfrak{A}$ and $\mathfrak{B}$ by $\mathfrak{A} + \mathfrak{B}$ and their least common multiple by $\mathfrak{A} - \mathfrak{B}$, then these two operations satisfy the conditions

(1) $\mathfrak{A} + \mathfrak{B} = \mathfrak{B} + \mathfrak{A}, \quad \mathfrak{A} - \mathfrak{B} = \mathfrak{B} - \mathfrak{A};$

(2) $(\mathfrak{A} + \mathfrak{B}) + \mathfrak{C} = \mathfrak{A} + (\mathfrak{B} + \mathfrak{C}); \quad (\mathfrak{A} - \mathfrak{B}) - \mathfrak{C} = \mathfrak{A} - (\mathfrak{B} - \mathfrak{C});$

(3) $\mathfrak{A} + (\mathfrak{A} - \mathfrak{B}) = \mathfrak{A}, \quad \mathfrak{A} - (\mathfrak{A} + \mathfrak{B}) = \mathfrak{A},$

which imply that

$$\mathfrak{A} + \mathfrak{A} = \mathfrak{A}, \quad \mathfrak{A} - \mathfrak{A} = \mathfrak{A}.$$

Then Dedekind introduces the following definition:

"If two operations on any two elements $\mathfrak{A}$ and $\mathfrak{B}$ of some (finite or infinite) system G always generate two elements $\mathfrak{A} \pm \mathfrak{B}$ such that conditions (1), (2), and (3) hold, then, regardless of the nature of these elements, the system G is called a dual group with respect to the operations $\pm$" (ibidem).[5]

As an example of a dual group Dedekind gives the system of modules that he investigated.[6] In another example he notes, in particular, that in a logical system one can interpret $\mathfrak{A} + \mathfrak{B}$ as a logical sum and $\mathfrak{A} - \mathfrak{B}$ as a logical product, and refers to Schröder's *Algebra of Logic*. Then he deduces properties of an arbitrary dual group.

We note that the most general definitions of a division ring and of a field were formulated as early as the end of the 19th century, and that a general abstract definition of a ring was given somewhat later (1910–1914) by Fraenkel and Steinitz. The final formulation of "modern algebra" was due to Emmy Noether (1882–1935) and her school. Generations of 20th-century mathematicians learned this subject from van der Waerden's famous *Modern Algebra*.

We have discused one series of investigations that have led to the creation of commutative algebra. Another series of investigations went on from the local methods initiated by Kummer and subsequently developed by Zolotarev and, especially, by Hensel. While Dedekind and Weber carried over the methods of study of fields of algebraic numbers to fields of algebraic functions, Hensel carried over series, the basic instrument for the representation and study of functions, to number theory. In 1897 he introduced p-adic and p-adic numbers.[7]

The ideas, methods, and concepts of commutative algebra evolved in the 19th century as a result of the parallel development and mutual influence of the theories of algebraic numbers and algebraic functions. These two areas were for mathematicians two models for the study of general laws. Progress in one area was carried over to the other area and stimulated its further development. The interaction of different areas of mathematics seems to be the mechanism of its development. At least, this is suggested by the example of the creation of commutative algebra, the basis and foundation of modern algebraic geometry.

We add a few words about the further investigations of fields of algebraic numbers. Here special attention was given to two issues. One was the question of unramified extensions $\mathbf{K}(\theta)$ of fields $\mathbf{K}$ of algebraic numbers, and the other was a proof of the most general reciprocity law in an arbitrary number field (Hilbert's ninth problem). The first of these issues was settled for extensions by Hilbert's class field theory (1899–1902), and the final solution of the second issue is due to Shafarevich (1948).[8]

Editor's notes

[1] One such example was given on p. 131.

[2] It should be pointed out that by requiring the system to contain infinitely many numbers Dedekind's definition excludes the ideal consisting of 0 alone, which is admissible under the modern definition.

[3] In this context, "integral function" means a function of Ω that satisfies a monic polynomial with coefficients in $\mathbf{C}[z]$. Such functions are said to be integral with respect to z. In a number field $\mathbf{K}$, the elements of $\mathbf{K}$ integral over the integers are the algebraic integers of $\mathbf{K}$. By analogy, functions integral with respect to z can be thought of as the "algebraic integers" of Ω. Thus "an ideal" refers to an ideal in the ring of functions of Ω integral with respect to z. Finally, note that this construction can be applied to any nonconstant $z \in \Omega$.

[4] One needs to assume that $\mathfrak{p}$ is a nonzero prime ideal. For Dedekind and Weber this was always true, but under the modern definition of ideal, $\{0\}$ is also prime in the rings considered here. For an example of the role played by the zero ideal, consider the dimension of a ring that is one less than the maximal length of a chain of prime ideals. The ring of functions of Ω integral with respect to z has dimension 1 because, if $\mathfrak{p}$ is a nonzero prime, it is maximal, and $\{0\} \subset \mathfrak{p}$ is a maximal chain. Similarly, the algebraic integers in a number field form a ring of dimension 1.

[5] Dual groups are closely related to Boolean algebras. In 1854 Boole wrote down for his (Boolean) algebras rules similar to Dedekind's rules (1), (2), (3).

[6] If $\mathfrak{A}$ and $\mathfrak{B}$ are ideals in the ring of algebraic integers in a number field, then their gcd is the sum $\mathfrak{A} + \mathfrak{B}$ and their lcm is the intersection $\mathfrak{A} \cap \mathfrak{B}$. It is easy to see that conditions (1), (2), and (3) hold in this case.

[7] This paragraph mentions a second series of investigations leading to commutative algebra, namely those stemming from local methods. It turns out that there is a third, equally important, series of investigations that contributed to commutative algebra, namely algebraic geometry, invariant theory, and the work of Kronecker and Hilbert. The rings considered by Dedekind and Weber (as discussed in the text) are all Dedekind domains, which means that all ideals are products of powers of distinct prime ideals (this is how unique factorization is recovered). But for other rings, such as $\mathbf{C}[x, y]$, this is no longer true—the ideal theory is much richer.

To see where such ideals might arise, first note that curves in the plane or surfaces in three dimensions have a single equation $F(x, y) = 0$ or $G(x, y, z) = 0$, so that ideals are not necessary. But what about the curve formed by the intersection of two surfaces $G_1(x, y, z) = 0$ and $G_2(x, y, z) =$

0? The polynomials G_1 and G_2 are not intrinsic to the curve, since other pairs of surfaces might intersect in the same curve. But the ideal generated by G_1 and G_2 is intrinsic. A similar example occurs in the plane. Suppose two given curves $F_1(x, y) = 0$ and $F_2(x, y) = 0$ intersect in a finite set of points. What can we say about a third curve $F(x, y) = 0$ that goes through the same points? In the simplest case, when the given curves intersect in smooth points with distinct tangents, the answer is that F must be of the form $AF_1 + BF_2$ for some polynomials A and B. In other words, F is in the ideal generated by F_1 and F_2. The general case, which involves certain multiplicity conditions, was proved by Max Noether in 1873 and was widely used in 19th-century algebraic geometry.

A more formal notion of this sort of ideal in a polynomial ring was given by Kronecker in 1882, where he used the term "Modulsystem". He wrote

$$G(x_1, x_2, \ldots, x_n) \equiv 0 \ (\mathrm{mod.}\ F_1, F_2, \ldots, F_n)$$

to say that G is in the ideal generated by $F_1, \ldots, F_n$, and he also explained what it meant for two sets of polynomials to generate the same ideal. Kronecker was especially interested in prime ideals, which generalize the irreducible polynomials so important in Galois theory. However, in contrast to Dedekind who considered an ideal as an infinite set (as we do today), Kronecker worked with finitely many elements at a time. Furthermore, Kronecker's main interest was divisors, which are a generalization (different from ideals) of Kummer's ideal numbers (see Harold M. Edwards, *Divisor Theory*, Birkhäuser, Boston, 1990).

In his great 1890 paper *Ueber die Theorie der algebraischen Formen*, Hilbert mentions explicitly both Noether and Kronecker and uses Kronecker's terminology for ideals, although his way of thinking of ideals was closer to Dedekind's than to Kronecker's. The theorems proved by Hilbert in this paper (Hilbert Basis Theorem and Hilbert Syzygy Theorem) are cornerstones of commutative algebra and algebraic geometry. He also introduced free resolutions, Hilbert functions, and Hilbert polynomials, which are important tools in modern commutative algebra. Another important discovery is Hilbert's Nullstellensatz from 1893, which describes the relation between the solutions of a system of equations $F_1 = \cdots = F_m = 0$ over the complex numbers and the ideal generated by $F_1, \ldots, F_m$. Hilbert proved all of these results in the course of his work on invariant theory.

But the question remained of the structure of ideals in a polynomial ring. This was solved by Lasker in 1905 with his discovery of primary decomposition for such ideals, which replaces the unique factorization of ideals in Dedekind domains. Trying to generalize Lasker's proofs is part of what led Emmy Noether to the discovery of Noetherian rings. These are rings in which every ideal has a finite generating set. She proved that every ideal in a Noetherian ring has a primary decomposition. Her drive to find the simplest conceptual basis for Lasker's theorems is part of what led to abstract algebra as we know it today.

[8] The last sentence of the chapter calls for two comments.

(a) Hilbert did not quite prove everything about what we now call the Hilbert class field. In particular, it was only in 1907 that his student Furtwängler clarified the role of the primes at infinity, and it was only in 1930 that Furtwängler (using the full power of class field theory) was able to prove the final piece, namely that all ideals of **K** become principal in the Hilbert class field.

(b) Many other people contributed to the solution of Hilbert's ninth problem. In particular, in 1920 Takagi was able to characterize all Abelian extensions of a number field, and somewhat later Artin proved the Artin reciprocity law, which provides an explicit link between Galois groups of Abelian extensions and generalized ideal class groups. (Artin proved his result for special cases in 1923, and more generally, using some important work of Chebotarev, in 1927.) Then Artin and Hasse worked out general reciprocity laws for lth powers (l prime) in 1925 and l^nth powers in 1928. This was completed by Shafarevich in 1950 with a general reciprocity law.

CHAPTER **9**

Linear and noncommutative algebra

1. Introduction of determinants

Thus far we have investigated two areas of the development of algebra in the 19th century. One of them—the theory of groups and fields—was connected with the algebraic solution of equations, and the other—commutative algebra—with number-theoretic investigations. These areas were very important but they were not the only ones. Determinants, matrices, quaternions, and algebras are some of the many concepts outside these areas that entered algebra in the 19th century. Their study gave rise to other concepts and methods.

We consider first the evolution of linear algebra. We are used to mentioning two of its key objects, determinants and matrices, in the same breath, but their historical origins are different, and they were far apart for a long time.

From the very beginning determinants were connected with the solution of systems of linear equations. In one of his letters to L'Hospital, Leibniz, who attached great importance to symbolism, wrote that he sometimes used numbers instead of letters. When L'Hospital expressed his puzzlement, Leibniz explained his method in a letter dated 28 April 1693. To solve the problem of elimination of the unknowns in the system

$$\begin{cases} a + bx + cy = 0, \\ d + ex + fy = 0, \\ g + hx + ky = 0, \end{cases}$$

he wrote it in the form

$$\begin{cases} 10 + 11x + 12y = 0, \\ 20 + 21x + 22y = 0, \\ 30 + 31x + 32y = 0. \end{cases} \tag{1}$$

149

Here the first entry in each coefficient denotes the number of the equation, and the second, the number of the constant term or of the unknown. This notation of Leibniz dates from 1684 and is the first instance of the double indexing of coefficients to which we are so used today.

The new notation enabled Leibniz to obtain certain general theorems and to express them in compact form. For example, a necessary condition for the solvability of the system (1) is the equality[1]

$$
\begin{aligned}
& 10 \cdot 21 \cdot 32 && 10 \cdot 22 \cdot 31 \\
&+11 \cdot 22 \cdot 30 \quad = \quad &&+11 \cdot 20 \cdot 32 \\
&+12 \cdot 20 \cdot 31 && +12 \cdot 21 \cdot 30
\end{aligned}
$$

This notation shows clearly that the second indices of the coefficients change cyclically.

Leibniz notes that: 1) the factors in each product come from different equations and different columns; and 2) the products have opposite signs if they have just one term in common (of course, this takes place only for third-order determinants).[2]

These investigations of Leibniz remained unknown. More than half a century later Gabriel Cramer (1704–1752) obtained results similar to but more general than those of Leibniz. In his "Introduction into the Analysis of Algebraic Curves" ("Introduction à l'Analyse des lignes courbes algébriques," 1750) he considered a system of n linear equations in n unknowns which he wrote as

$$
A^1 = Z^1 z + Y^1 y + X^1 x + V^1 v + \cdots,
$$
$$
A^2 = Z^2 z + Y^2 y + X^2 x + V^2 v + \cdots,
$$

$$\cdots\cdots\cdots\cdots\cdots\cdots\cdots\cdots\cdots\cdots\cdots\cdots$$

Cramer expressed all unknowns as fractions with the same denominator —the sum of products of the form $\pm ZYXV \ldots$. The letters $Z, Y, X, V, \ldots$ are taken with indices obtained from the $n!$ permutations of the numbers $1, 2, \ldots, n$. To determine the sign of a product Cramer introduced the notion of a "disorder" (dérangement) in the position of indices: a "disorder" (i.e., an inversion) occurs if a larger number precedes a smaller one. The product is taken with a plus or minus sign depending on whether the number of "disorders" in a permutation is even or odd. In modern terms, the denominator is equal to the determinant

$$\begin{vmatrix} Z^1 & Y^1 & X^1 & V^1 & \dots \\ Z^2 & Y^2 & X^2 & V^2 & \dots \\ \vdots & \vdots & \vdots & \vdots & \end{vmatrix}$$

The numerators in the expressions for the unknowns are obtained from the denominators by replacing the coefficients of an unknown by the constant terms with the same indices.

Cramer's rule rapidly gained wide currency. This was due in part to its use by Bézout for the elimination of one of the unknowns from a pair of equations $f_1(x, y) = 0$ and $f_2(x, y) = 0$, where f_1 and f_2 are polynomials.

Gradually determinants themselves became an object of study. In particular, they were investigated by A. T. Vandermonde (1735–1796) and by Laplace. An exhaustive theory of determinants was developed by Cauchy in a memoir of 1815. He introduced the term "determinant", used hitherto by Gauss to denote the discriminant of a quadratic form in two or three unknowns (1801).

Further progress in the study of systems of linear equations, in particular the determination of conditions for their consistency and for the uniqueness of their solutions, was achieved much later, namely when the study of determinants was combined with the study of matrices.

2. Linear transformations and matrices

The origin of matrices was very different from that of determinants. From the very beginning they were introduced as an abbreviated description of a linear transformation. The first instance of such a description of a linear transformation by a table of coefficients is found in Gauss' *Disquisitiones*. In Chapter V he considers a ternary quadratic form

$$f = ax^2 + a'y^2 + a''z^2 + 2bxy + 2b'xz + 2b''yz$$

with integral coefficients. As in the case of a binary quadratic form, he makes a substitution S

$$x = \alpha x' + \beta y' + \gamma z',$$
$$y = \alpha' x' + \beta' y' + \gamma' z',$$
$$z = \alpha'' x' + \beta'' y' + \gamma'' z',$$

with integral coefficients. This substitution takes the form f into a form g.

Then, writes Gauss, we will, for brevity, ignore the variables and say that f goes over into g under the substitution

$$\begin{array}{ccc} \alpha & \beta & \gamma \\ \alpha' & \beta' & \gamma' \\ \alpha'' & \beta'' & \gamma'' \end{array} \qquad \text{(S)}$$

(Gauss, *Disquisitiones*, Section 268, p. 294 of the English translation). Continuing, Gauss states that if

f goes over into f' under the substitution			f' goes over into f'' under the substitution		
α	β	γ	δ	ϵ	ξ
α'	β'	γ'	δ'	ϵ'	ξ'
α''	β''	γ''	δ''	ϵ''	ξ''

then f goes over into f'' under the substitution

$$\begin{array}{ccc} \alpha\delta + \beta\delta' + \gamma\delta'' & \alpha\epsilon + \beta\epsilon' + \gamma\epsilon'' & \alpha\zeta + \beta\zeta' + \gamma\zeta'' \\ \alpha'\delta + \beta'\delta' + \gamma'\delta'' & \alpha'\epsilon + \beta'\epsilon' + \gamma'\epsilon'' & \alpha'\zeta + \beta'\zeta' + \gamma'\zeta'' \\ \alpha''\delta + \beta''\delta' + \gamma''\delta'' & \alpha''\epsilon + \beta''\epsilon' + \gamma''\epsilon'' & \alpha''\zeta + \beta''\zeta' + \gamma''\zeta'' \end{array}$$

Thus he gave the rule for multiplication of 3×3 matrices.

Gauss also introduced the concept of the transpose

$$\begin{array}{ccc} \alpha & \alpha' & \alpha'' \\ \beta & \beta' & \beta'' \\ \gamma & \gamma' & \gamma'' \end{array} \qquad \text{(S'')}$$

of a substitution. He wrote: "We will say that the substitution (S'') is obtained from the substitution (S) by transposition."

We note that in his memoir of 1815 Cauchy gave a rule for the multiplication of determinants based on Gauss' rule for the multiplication of matrices.

The introduction of the notion of n-dimensional space in the 1840s marked a turning point in the evolution of linear algebra. The term itself first appeared in the title of Cayley's paper "Chapters of Analytic Geometry of n-Dimensions" (1843). Everything in this paper was done in a purely algebraic way but towards the end Cayley applied his results to "the case of four variables." The application shows that he treated the variables $x_1, \ldots, x_n$ as projective coordinates of a point in $(n - 1)$-dimensional projective space. He stressed that the geometric terminology was just a convenient language and that a vector in n-dimensional space was merely an ordered n-tuple $(x_1, \ldots, x_n)$ of real numbers.

Hermann Grassmann's famous *Doctrine of Linear Extension* (*Lineare Ausdehnungslehre*) appeared in 1844. This work was far more profound than Cayley's. According to Grassmann, it was necessary to introduce into geometry "oriented magnitudes" of arbitrary dimension. Vectors that were differences of two points (beginning and end) represented first-level magnitudes (erste Stufe). The product of two such vectors was to be an oriented two-dimensional magnitude, and so on. But in his book Grassmann failed to give sharp definitions and tended to resort to philosophical arguments. As a result, in spite of the book's rich content, contemporary mathematicians failed to understand his ideas. In 1862 Grassmann published a more "mathematical" version of his book, titled *Doctrine of Extension* (*Ausdehnungslehre*), in which he considered various types of multiplication of vectors (scalar, vector, and others), defined linear independence, and introduced "extensive magnitudes" (now called nth order Grassmann numbers) and their outer products. Nonetheless, Cayley's viewpoint remained dominant practically until the 1930s.

Ludwig Schläfli's (1814–1895) book on multidimensional geometry appeared in 1851, and in 1854 Berhard Riemann, one of the greatest mathematicians of the 19th century, introduced in his famous habilitation lecture "On the Hypotheses that Lie at the Foundation of Geometry" the concept of an n-dimensional manifold and used Gauss' method of constructing the geometry of a surface to construct n-dimensional geometries.

According to Klein, "around 1870 the concept of a space of n dimensions became the general property of the advancing young generation". But even earlier this concept played a significant role in linear algebra. Cayley's famous "A Memoir on the Theory of Matrices" (Philosophical Transactions, 1858) appeared in 1853. In this memoir Cayley introduced "matrices", $m \times n$ arrays of the coefficients of a transformation, which served as compact descriptions of linear transformations. For square matrices ($m = n$) Cayley developed a calculus that involved the operations of addition and multiplication (the latter for an $n \times m$ and an $m \times k$ matrix). The product of two matrices was the matrix of the transformation that was the result of the successive application of the linear transformations corresponding to the two matrices. Cayley verified the associativity and noncommutativity of matrix multiplication, introduced a unit matrix and a zero matrix, and explained the connection between a square matrix and its determinant. For the latter he introduced the modern notation

$|a_{ij}|$. Finally, he showed that the determinant of a product of two matrices is the product of the determinants of the factors.

Generalizing one of Hamilton's theorems on quaternions, Cayley stated the theorem that every square matrix satisfies its characteristic polynomial (the Cayley–Hamilton theorem) and proved it for matrices of order two and three. The important point here is that Cayley was considering matrix equations, and in so doing he was extending the realm of algebra from numbers to objects very different from numbers (in modern terms, square matrices of a given order form an algebra but not a field).

In the same memoir Cayley determined the first linear representation of an algebra. Specifically, he showed that the algebra of quaternions is isomorphic to the algebra of second-order matrices

$$\begin{pmatrix} a + di & b + ci \\ -b + ci & a - di \end{pmatrix} \tag{2}$$

in the sense that both have the same multiplication tables.

What was missing in Cayley's memoir was the notion of rank of a matrix. This concept, and its application to the study of the solvability of a system of linear equations (the so called Kronecker–Capelli theorem), were discovered independently by a number of authors. The first to publish this theorem was Charles L. Dodgson (known under the pseudonym of Lewis Carroll) (1832–1898).

The problem of reduction of a matrix to canonical form was solved independently by Karl Weierstrass (1815–1897) and by Jordan. Weierstrass used elementary divisors to give a necessary and sufficient condition for similarity of matrices. In his *Treatise on Substitutions* Jordan proved the possibility and uniqueness of the reduction of a matrix to normal form.

Already Euler considered the problem of reduction of a symmetric quadratic form $\sum a_{ij}x_ix_j$, $a_{ij} = a_{ji}$, to a sum of squares. He posed it in connection with the problem of classification of quadratic curves and surfaces by reducing them to canonical form. Around 1850, Jacobi and Sylvester independently solved the problem by establishing the so-called law of inertia of quadratic forms.

We see that all the fundamental theorems of linear algebra were proved by the 1870s. By that time the notion of an n-dimensional space entered mathematical practice. It is well known that the methods of linear algebra were immediately and extensively used in all areas of mathematics, both pure and applied. The range of application of these methods continues to grow.

3. The English school of symbolic algebra.
Hamilton's quaternions

In the 18th century English mathematicians adhered dogmatically to the New-
tonian tradition (to the point of rejecting the convenient differential notation
of Leibniz) and ended up in the margin of evolving continental mathematics.
At the beginning of the 19th century a group of young Cambridge mathematicians, who called themselves the Analytical Society, decided to abandon this
tradition. They adopted the Leibniz notation in analysis and joined mainstream
European mathematics. They paid special attention to the role of symbolism
and created a new conception of algebra. George Peacock, one of the founders
of the Analytical Society, divided algebra into two parts: the numerical and the
symbolic. The first operates with letters that always stand for numbers and the
second with pure symbols that need no interpretation. For these symbols one
can introduce algebraic operations with properties prescribed *a priori*, i.e.,
axiomatically. But the laws governing these operations must coincide with
the rules of numerical algebra in the special case when the symbols stand for
ordinary numbers (principle of permanence).

What Peacock and others were unaware of was that some choices of
axioms or rules of operation are inconsistent and therefore cannot be realized
for actual mathematical objects. For example, it was initially assumed that it is
possible to introduce numbers depending on three or more linearly independent
units,

$$\alpha = a_1 l_1 + \cdots + a_n l_n, \quad n \geq 3,$$

and define for them operations of addition, subtraction, multiplication, and
division satisfying the same properties as in the case of real numbers. This
eventually proved to be impossible. Already Augustus De Morgan and Duncan
F. Gregory, representatives of the school of symbolic algebra, tried in vain
to introduce such numbers for $n = 3$, but it took time to understand the
insurmountable limitations in the conceptions of symbolic algebra.

In the meantime a number of English mathematicians who adopted the
basic tenets of symbolic algebra obtained new and important results. Some of
the results, namely Cayley's axiomatic definition of a group and his matrix
calculus, were discussed earlier (see Chapter VII). At this point we mention
George Boole's (1815–1864) algebra of logic with the following rules of
operation (or axioms):

$$x(1 - x) = 0,$$

$$xy = yx,$$

$$xx = x, \quad x + x = x,$$

$$x + y = y + x,$$

$$x(u + v) = xu + xv.$$

Here x and y are to be thought of as classes of objects, 1 as the universal class, multiplication as intersection of two classes, and addition as their union. The law of the excluded middle can be written as

$$x + (1 - x) = 1.$$

One of the people influenced by ideas of symbolic algebra was the famous Irish mathematician William Rowan Hamilton (1805–1865). He began with Gauss' complex numbers.

As early as 1835 Hamilton gave a purely arithmetical interpretation of complex numbers. He considered ordered pairs (a, b) of real numbers and identified the pair $(a, 0)$ with the real number a. He called $(1, 0)$ the first unit and $(0, 1)$ the second unit. He regarded two pairs (a, b) and (c, d) as equal if and only if $a = c$ and $b = d$. His definitions of addition and subtraction were the obvious ones:

$$(a, b) \pm (c, d) = (a \pm c, b \pm d).$$

As a first step towards a definition of multiplication Hamilton put

$$(0, a)\,(0, b) = (\gamma_1\, ab,\ \gamma_2\, ab).$$

He noted that it is convenient to choose $\gamma_1 = -1$ and $\gamma_2 = 0$ (but that these constants can also be chosen differently). These choices, and the distributive law of multiplication over addition, imply that

$$(a, b)\,(c, d) = [(a, 0) + (0, b)]\,[(c, 0) + (0, d)] = (ac - bd, ad + bc).$$

In particular,

$$(0, 1)^2 = (0, 1)\,(0, 1) = (-1, 0),$$

so that

$$\sqrt{(-1, 0)} = \sqrt{-1} = (0, 1).$$

In the same year, i.e., in 1835, Hamilton tried to generalize complex numbers by introducing three units $(1, 0, 0)$, $(1, 0, 0)$, and $(0, 0, 1)$. We note that this approach was essentially different from the one adopted in continental Europe which led to the introduction of algebraic numbers. Hamilton adopted three new linearly independent units and considered ordered triples (a, b, c), briefly triplets, of real numbers. He defined addition and subtraction coordinatewise. He wanted to define multiplication so that the product of two triplets was again a triplet,

$$(a_1, b_1, c_1)(a_2, b_2, c_2) = (a_3, b_3, c_3),$$

and so that the following two conditions were satisfied:

1. Multiplication of triplets is to be distributive over addition (i.e., triplets can be multiplied like polynomials).

2. Define the modulus of (a, b, c) to be $a^2 + b^2 + c^2$ (i.e., the square of the length of the corresponding vector). Then the modulus of the product of two triplets is to be equal to the product of the moduli of the factors. Hamilton called this condition the law of moduli.

Now we know that for these requirements to be satisfied the number of basic units must be 1, 2, 4, 8, or 16 (a result discovered by Adolf Hurwitz in 1898). In other words, Hamilton's attempt was doomed to failure. We can reconstruct his reasoning from his notebooks and from his letters to John Graves and to his son. (Such a reconstruction is found in [8]. Hamilton described his 1835 investigation of triplets in the introduction to his *Lectures on Quaternions* of 1853.) Putting $i^2 = j^2 = -1$ and $ij = ji$, he set down the following rule of multiplication of triplets

$$(a+bi+cj)(x+yi+zj) = (ax-by-cz)+i(ay+bx)+j(az+cx)+ij(bz+cy).$$

The problem was how to define the product ij. Hamilton assumed that $ij = \alpha + \beta i + \gamma j$, i.e., that ij is again a triplet. But all his attempts to determine the values of α, β, and γ so as to satisfy conditions 1 and 2 ended in failure. He tried and rejected the possibilities $ij = 1$, $ij = -1$, $ij = 0$, and $ij = ji$. Finally, as he puts it himself, on Monday, October 16, 1843, a flash of insight befell him. He realized that it was necessary to put $ij = k$, where k is a new imaginary unit not expressible in terms of i, j, 1. As van der Waerden puts it, Hamilton made "a leap into the fourth dimension" and began to consider quaternions

$$a + bi + cj + dk,$$

briefly (a, b, c, d), for which he established the following rules of multiplication:

$$ik = iij = -j; \quad kj = ijj = -i; \quad ki = j; \quad jk = i.$$

And, finally, $k^2 = ijij = -iijj = -1$.

In a letter to his son, written shortly before his death, Hamilton says that the flash of insight occurred as he was walking to attend and preside over a meeting of the Royal Irish Academy: "...your mother was walking with me along the Royal Canal...; and although she talked with me now and then, yet an undercurrent of thought was going on in my mind, which gave at last a result..." Hamilton immediately set down his discovery in his notebook and cut the laws of multiplication of quaternions with a knife on a stone of a bridge. On the next day he wrote about his discovery to John Graves.

Hamilton's rule of multiplication of quaternions is designed so that every nonzero quaternion has an inverse. Indeed, let $q = a + bi + cj + dk$. By analogy with the complex numbers we introduce its "conjugate" $\bar{q} = a - bi - cj - dk$. It is easy to see that $q\bar{q} = a^2 + b^2 + c^2 + d^2 = N(q)$ (the "norm" of q). It follows that $q^{-1} = \bar{q}/N(q)$.

This is the story of the introduction into mathematics of the first non-commutative division algebra.

Hamilton and his students were so enthralled by the theory of quaternions that they tried to base on it all of analysis and classical mechanics. In time mathematicians abandoned this idea. But today quaternions turn out to be extremely useful in quantum mechanics.[3]

We conclude by noting that Gauss had discovered the law of multiplication of quaternions before 1820. Like so many of his other discoveries, this one was not published in his lifetime.

4. Algebras

Hamilton's ideas, contained in a letter to John Graves dated October 17, 1843, fell on fertile soil. Already in December of that year Graves constructed an algebra with eight basis elements. In 1845, independently of Graves, Cayley also constructed such an algebra. Numbers in these systems came to be known as Cayley numbers or octonions. Here multiplication is not only noncommutative but also nonassociative.

Another generalization is due to Hamilton who tried to introduce quaternions over the field $\mathbf{C}$ of complex numbers. But here he was in for a surprise: it turned out that this system contains divisors of zero. Indeed, over the field $\mathbf{C}$

$$x^2 + 1 = (x + i)(x - i).$$

Substituting the unit j for x we obtain

$$(j + i)(j - i) = j^2 + 1 = 0.$$

Subsequently, algebras of finite dimension were investigated by Benjamin Peirce (1809–1880), Georg Frobenius (1849–1917), and Fedor E. Molin (1869–1951).

The field of complex numbers, the skew field of quaternions, and matrix algebras are instances of associative algebras. A general definition of an associative algebra was given by Peirce in his *Linear Associative Algebras* (1872). Peirce defined such an algebra as an n-dimensional vector space on which there is defined an associative multiplication that is distributive over (the vector space) addition. If $e_1, e_2, \ldots, e_n$ are the basic units (or basis vectors), then the multiplication is determined as soon as we are given the multiplication table $e_i e_j = \sum c_{ij}^k e_k$ of the basis vectors (the structural formulas of the algebra), with the structural constants c_{ij}^k chosen so that $e_i(e_j e_k) = (e_i e_j)e_k$.

Peirce introduced the notions of a *nilpotent* element e such that $e^r = 0$ and of an *idempotent* element e such that $e^2 = e$.

The special position of the field of complex numbers and of the skew field of quaternions in the class of associative algebras is apparent from the Weierstrass–Frobenius theorem, which asserts that the only associative division algebras of finite dimension over the field $\mathbf{R}$ of reals are the skew field of quaternions, the field of complex numbers, and the field $\mathbf{R}$ itself.

This theorem was proved by Frobenius in 1878 and two years later, independently of Frobenius, by Charles Peirce (1839–1914). The reason for attaching Weierstrass' name to the theorem is that he proved earlier an important special case, namely, that the only commutative algebras satisfying the conditions of the theorem are $\mathbf{C}$ and $\mathbf{R}$. This showed that, far from being an exception, the occurrence of divisors of zero is rather the rule.

Weierstrass began to investigate the structure of commutative algebras in 1861 and presented his results in his lectures (his results were published in 1884 in the paper "On the Theory of Complex Quantities Formed from n Principal Units" (*Gött. Nachr.* 1884)). He gave a general definition of an algebra of finite dimension and introduced the important concept of the direct sum of algebras $\{\sum a_i e_i\}$ and $\{\sum b_j f_j\}$ as the algebra

$$C = A \oplus B = \left\{\sum a_i e_i\right\} \oplus \left\{\sum b_j f_j\right\}, \quad e_i f_j = f_j e_i = 0.$$

He also defined the notion of a nilpotent element and showed that every commutative algebra without nilpotent elements (over $\mathbf{C}$ or $\mathbf{R}$) is the direct sum of a number of fields isomorphic to $\mathbf{C}$ or $\mathbf{R}$. A generalization of this result is known as the Wedderburn theorem.

Some important results in the theory of algebras were obtained at the turn of the 20th century by the eminent Russian mathematician Fedor E. Molin.

A graduate of Dorpat (now Tartu) University, Molin worked first at Dorpat and then at Tomsk. He was the first to give (in 1903) a precise definition of the concept of a *radical* (the term "radical" is due to Frobenius). Also, he introduced the notion of a *factor algebra* of an algebra (which corresponds to the notion of a two-sided ideal), and, by analogy with simple groups, defined a *simple algebra* (of finite dimension) as an algebra without nontrivial two-sided ideals. His fundamental result is that every simple algebra (other than C) is a full matrix ring of order n whose elements are either real numbers (an algebra of dimension n^2), or complex numbers (an algebra of dimension $2n^2$), or quaternions (an algebra of dimension $4n^2$).[4]

Further fundamental investigations of Molin, as well as of Frobenius, supplemented later by Elie Cartan, pertain to the structure of semisimple algebras (i.e., algebras without a radical).[5]

We cannot here go into the history of algebras in the 20th century other than say that this branch of mathematics continues to flourish today. There is no doubt that what stimulated its rise was Hamilton's introduction of quaternions.

Editor's notes

[1] If we homogenize (1) by attaching a third variable to the constant terms, then this equality is a necessary and sufficient condition for the homogenized system to have a nontrivial solution.

[2] In addition to Leibniz, Maclaurin (1748) also used determinants (in poor notation) to solve systems of linear equations in two, three, and four variables.

[3] Modern quantum mechanics uses Clifford algebras to discuss spin. In the simplest case, the Pauli spin matrices lead naturally to a subalgebra of a certain Clifford algebra which consists of the matrices in (2), and hence via (2) to the quaternions.

[4] Molin did some extremely influential work on algebras and introduced some important concepts, but the most important results belong to Wedderburn. For example, the result just quoted is a special case of the Wedderburn Structure Theorem, which is proved in standard books on associative algebras.

[5] These results are also contained in the Wedderburn Structure Theorem mentioned above. More details on the work of Molin, Frobenius, Cartan, and Wedderburn can be found in [9].

CONCLUSION

We have studied the evolution of some, but certainly not all, branches of algebra. (An example of an omitted theory is the theory of invariants.) We hope that our incomplete survey has given the reader some idea of how this discipline, which began some four thousand years ago with observations on the laws of arithmetical operations and their first applications, became ever more abstract, developed its proper language in the form of a literal calculus, and eventually ended up with an extraordinary trove of ideas, methods, and theories with which it can now investigate the most general systems of objects (far removed from numbers) on which there are defined binary operations with extremely varied properties.

At first algebra developed very slowly. Some literal notation appeared only in the the third century CE (Diophantus) and was limited to a single variable and some of its powers. The introduction of literal parameters, that is, the use of letters for denoting coefficients, came only in the 16th century, and thus more than 13 centuries later. In Diophantus' *Arithmetic* we see the first records of algebraic equations and the first extension of the number system (from whole numbers to the field of all positive and negative rational numbers). Complex numbers too appeared only in the 16th century, and the rigorous introduction of the real numbers took place only in the 1870s, that is, some 300 years later.

An explosive development of algebra occurred in the 19th century as a result of the introduction of a multiplicity of new and very abstract objects, such as groups, rings, fields, ideals, matrices, algebras, and so on, and of the development of new methods for their study. At first glance it might seem that algebra migrated to a domain of very high abstractions far removed from the real world and from its science. But this is a deceptive impression. It was abstractions of a higher order, and the construction of new abstract theories

161

by new methods, that enabled the algebra of the last two centuries to penetrate so deeply into the quantitative regularities of the real world that it could be extensively applied not only in all areas of mathematics but also in physics (and especially in quantum mechanics).

A new view of the geometric algebra
of the ancients*

In the last twenty years a discussion of the "geometric algebra" of the ancient Greeks has taken place in the history of mathematics. The key question which has attracted the attention of students of the mathematics of ancient civilizations has been whether "geometric algebra" was just a speculative construct of H. G. Zeuthen [8] and P. Tannery [16], who introduced modern notions into ancient mathematics, or whether these scholars managed to retrieve the true essence of the creations of the great Greeks.

The participants in this discussion included B. L. van der Waerden [18], H. Freudenthal [12], S. Unguru [17], and other historians of science. In spite of van der Waerden's penetrating analysis, persuasive logic, and numerous examples in favor of the argument of the existence of geometric algebra, each of the participants in the discussion stuck to his own views. Convinced of the futility of this approach, we have decided to look at the problem from the position of the opponents of modernization in the history of science, i.e., to consider the problem in the context of the science of the ancient world.

We propose to show that geometric language was adopted for the presentation and justification of the elements of algebra not only in ancient Greece but also in the ancient civilizations of China, India, and very likely, Babylon. Moreover, geometry was the only possible universal language of antiquity. This same language continued to serve the science of the Middle Ages in the Arab East and in Europe. It was only at the end of the 16th century that this language was replaced by the new language of the literal calculus, and,

---

* *Translator's note.* The following paper by the authors of this book is an account of their new view of the rise and evolution of Greek geometric algebra. It differs from the "standard" account reflected in the last paragraph of §1 in Chapter II and is included here at their request.

to a large extent, this development determined the structure and character of post-Renaissance mathematics.

Before we can answer the question about the existence of geometric algebra we must decide what this term is to mean. This brings us in a natural way to a second, more difficult, question, namely, what do we mean by algebra, and when is it legitimate to talk about the presence of algebraic elements.

If we say that algebra began when literal calculus was created and it became possible to operate with formulas, then the family tree of algebra goes back to the works of Viète, i.e., to the end of the 16th century. This was the view of the famous Soviet mathematician B. N. Delone (1890–1980). One can relax this requirement somewhat and say that the elements of algebra came into being with the first use of symbolism. Then the creator of this science will have been Diophantus of Alexandria, and the time of its birth, the 3rd century CE. One could also adopt the opposite view and say that algebra came into being only when mathematicians began to investigate algebraic structures such as groups, rings, fields, ideals, and so on. Then the birth of algebra would have to be moved up to the 19th century, for up to that time algebra was just "the art of solving equations".

But one may also regard as the elements of algebra the investigation and solution of the class of problems associated with algebra today using methods that are likewise so associated; i.e., the establishing of basic algebraic identities and the formulation and solution of problems equivalent to quadratic and cubic equations. In that case, the beginnings of algebra would have to be moved back to the 2nd millennium BCE. We propose to adopt the latter point of view. It now remains to agree to the meaning of the term "geometric algebra".

Whenever the solution of an algebraic problem is carried out by means of geometric diagrams and geometric constructions we will say that the problem has been solved by methods of geometric algebra, i.e., that elements of algebra have been dealt with using the language of geometry.

Having thus defined geometric algebra, we can expect that the first step on the road of translation of algebraic concepts into the language of geometry will be the identification of magnitude (a given—a number or an unknown) with some geometric image.

In the ancient Indian *Sulvasūtras* (7–5 centuries BCE), as well as in classical Greek geometric algebra, every magnitude is represented by a segment. The product xy of two magnitudes is a rectangle on the corresponding segments, or an "area"; and x^2 and y^2 are squares with sides x and y respectively. In ancient China identity transformations were always applied to magnitudes associated with some geometric images. While it is true that the ancient Babylonians do not seem to have used geometric diagrams to justify identities and to solve equations, the use of terms such as "length", "width", and "area" for the designation of unknowns suggests that geometric interpretations were used to some extent in ancient Babylon as well.

We give examples which substantiate our claims. The algebraic identity

$$(a + b)^2 = a^2 + b^2 + 2ab, \tag{1}$$

which was known in all ancient civilizations, becomes obvious when translated into the language of geometry

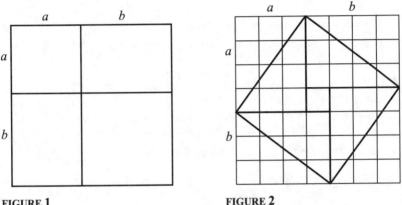

FIGURE 1   **FIGURE 2**

Figure 1 represents the ancient Greek and Indian versions of such a translation.

The ancient Chinese variant of the proof of the identity (1) can be easily obtained from the well-known "hypotenuse drawing" in the *Mathematical Treatise on zhou be* (Figure 2), one of the oldest sources of Chinese mathematics that has come down to us. Most scholars are of the opinion that Chao Chun-Ch'ing's (3rd century CE) extensive commentary on this work contains one of the earliest proofs of Pythagoras' theorem. We will show below that this proof was carried out using geometric algebra.

Both drawings not only prove (1) but also have an intuitive appeal. This is undoubtedly one of the distinctions of this method of establishing truths. We now present the ancient Chinese proof of Pythagoras' theorem as presented in the *Mathematical Treatise on zhou be* (Figure 3).

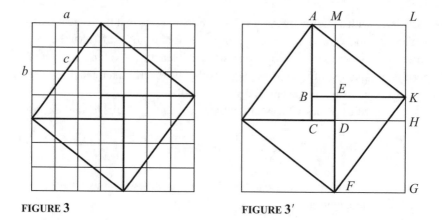

FIGURE 3   FIGURE 3′

It is easy to see that the large square on side $a + b$ is made up of the square on side c and of the four right triangles with sides a and b, so that

$$(a + b)^2 = c^2 + 2ab.$$

In view of (1), we find that $c^2 = a^2 + b^2$.

This variant of the reconstruction of the proof is due to van der Waerden [19]. B. Gillon suggests another method of proof. If in Figure 3 we denote points by letters as in Figure 3′, then we have

$$(a-b)^2 + 2ab = ALKB + EFGK + CBDE = ACHL + DHFG = a^2 + b^2.$$

Since Chao Chun Ch'ing's commentary contains a proof of the fact that $c^2 = (a - b)^2 + 2ab$, Pythagoras' theorem follows.

According to the author of the treatise, Pythagoras' theorem was known to the ancient Chinese as early as the 6th century BCE, and they knew about the 3, 4, 5 right triangle as early as the 12th century BCE. Here is the relevant ancient text:

"If in an angle bar the width is 3—side hou—and the length is 4—side hu—then the transversal [which connects the ends of the angle] is 5" [9, p. 14].

As indicated by the treatise, Chougun Dan' heard this ancient formulation from the high official Shan Gao who, in turn, refers to an even earlier time when the legendary Fusi (3rd millennium BCE) ruled "the Skies by means of numbers" [ibidem].

The Indian *Sulvasūtras* contain a diagram similar to that in Figure 3. It is easy to see that if we accept van der Waerden's reconstruction of the proof of Pythagoras' theorem in ancient China, then this diagram can also be used to obtain the algebraic identity

$$(a - b)^2 = a^2 + b^2 - 2ab. \tag{2}$$

Indeed, the square on side c is made up of the square on side $a - b$ and of four right triangles with sides a and b. Hence

$$(a - b)^2 = c^2 - 2ab.$$

Since $c^2 = a^2 + b^2$, (2) follows.

Thus the same diagram could be used to prove Pythagoras' theorem, to establish the algebraic identities $(a = b)^2 = a^2 + b^2 + 2ab$ and $a^2 - b^2 = (a - b)(a + b)$ and, as we will show below, to solve quadratic equations.

As mentioned earlier, one of the basic classes of problems studied by ancient mathematicians was the solution of problems equivalent to quadratic equations (that is why Renaissance mathematicians, following Arab mathematicians headed by al-Khwarizmi, called algebra "the art of simplification and solution of equations"). We will see how such problems were solved in the ancient sources.

In the ancient Indian *Sulvasūtras* the quadratic equation $x^2 = ab$ is formulated as the problem: "Transform a rectangle into a square", and its solution is given by the construction represented in Figure 4.

Let $AB = a$, $AD = b$. Construct the segment AE equal to b. Halve $BE : BC = CE = (a - b)/2$. Apply the rectangle $FDML = CEFH = BCHG$ to the segment FD. This transforms the rectangle $ABCD$ into the

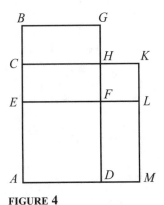

FIGURE 4

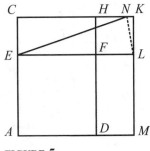

FIGURE 5

gnomon $ACHFLM$. Hence

$$ab = \left(\frac{a+b}{2}\right)^2 - \left(\frac{a-b}{2}\right)^2. \qquad (3)$$

Now the side x of the required square can be found by means of the construction represented in Figure 5.

Here $AC = (a+b)/2 = AM$, $CE = (a-b)/2$. Construct the segment $NE = LE = AC$. Since $CN^2 = NE^2 - CE^2$, the segment $CN = x$. Thus to solve a pure quadratic equation one must have the identity (3). Figure 4 can be viewed as its geometric proof.

Euclid formulates this problem as follows: "Construct a square equal to a given rectangle". To within a rotation, Figure 4 is found in proposition 6 of Book II of his *Elements* (we will refer to it in the sequel as II$_6$). A similar construction, given by Euclid in II$_5$, can also be used to justify the identity (3). Specifically, put $AB = a$, $BD = b$. In Figure 6,

$$AC = CD = (a+b)/2, \quad BC = (a-b)/2, \quad AB \cdot BD = CD^2 - CB^2. \quad (4)$$

In Figure 7,

$$AC = CD = (a-b)/2, \quad BC = (a+b)/2, \quad AB \cdot BD = CB^2 - CD^2. \quad (5)$$

However, as was shown by Zeuthen [8], the main reason for the appearance of propositions 5 and 6 in Euclid is not the proof of the identity (3) but the solution of two types of quadratic equations:

$$\text{the elliptic type} \qquad x(a - x) = S, \qquad (6)$$

$$\text{the hyperbolic type} \qquad x(a + x) = S. \qquad (7)$$

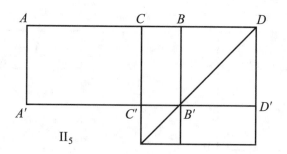

II$_5$

FIGURE 6

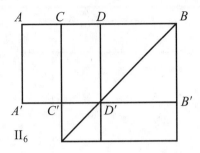

FIGURE 7

The ancient mathematicians formulated the corresponding problems geometrically in terms of "application of areas". The problem corresponding to equation (6) is stated as follows: "apply" the area S to the segment a so that the "deficiency" is a square. Then in Figure 6, $AD = a$, $ABB'A' = S$, and the "deficiency" $BDD'B'$ is the required square.

Similarly, in the problem corresponding to equation (7) one had to "apply" the area S to the segment a so that the "excess" was a square (Figure 7). Then equations (6) and (7) can be rewritten, respectively, as

$$S = x(a - x) = (a/2)^2 - (a/2 - x)^2,$$

and

$$S = x(a + x) = (a/2 + x)^2 - (a/2)^2.$$

In both cases the rectilinear area S is represented as the difference of squares. Thus in order to solve an equation of type (6) or (7) one must first transform S into a square and then use Pythagoras' theorem to find $a - x$ in case of (6) and $a + x$ in case of (7).

The very same two types of quadratic equations with their respective geometric solutions can also be found in the *Mathematical Treatise on zhou be* or, more accurately, in Chao Chun-Ch'ing's commentary on this treatise. He first gives a geometric justification (Figures 8 and 9 below) of the two symmetric identities

$$c^2 - b^2 = (c - b)(c + b) \quad \text{and} \quad c^2 - a^2 = (c - a)(c + a),$$

where $a, b,$ and c are linked by Pythagoras' theorem (here, and throughout the treatise, $a, b,$ and c are the sides and hypotenuse of a right triangle, i.e., $c^2 = a^2 + b^2$):

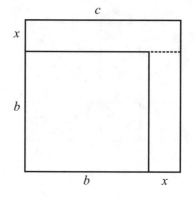

The area of the gnomon-like strip in Figure 8, referred to in the text as the "angle bar from the dividend for hou", is a^2 and consists of two rectangles with respective areas cx and bx, where $x = c-b$. Hence $cx+bx = x(c+b) = a^2$, or

$$a^2 = (c - b)(c + b),$$

so that

$$c^2 - b^2 = (c - b)(c + b).$$

Similar computations applied to Figure 9 yield

$$c^2 - a^2 = b^2 = (c - a)(c + a).$$

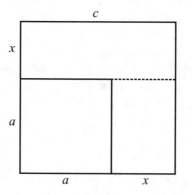

FIGURE 9

As pointed out by the Chinese historian of science Tsian' Bao-tsum [9], we have here a geometric solution of a pair of quadratic equations

$$x^2 + 2bx = a^2 \quad \text{and} \quad x^2 + 2ax = b^2,$$

whose respective roots are $c - b$ and $c - a$.

We see that in all known works of ancient mathematicians one very frequently comes across geometric terminology in connection with the formulation of algebraic problems and that solutions are sought by means of corresponding geometric constructions. This tradition has been maintained until very late. We find geometric constructions meant to illustrate the solutions of algebraic equations in the works of Arabic mathematicians in the Middle Ages. In the works of the great Renaissance mathematicians Cardano, Bombelli, and others, every analytic argument bearing on the determination of the roots of linear, quadratic, and cubic equations is accompanied by corresponding geometric constructions. And, even long after the creation of the literal calculus, post-Renaissance mathematicians continued to apply geometric interpretations to propositions obtained by analytic means.

In our view, this confirms the conclusion that for a long time only geometry could be the language of mathematics, for only geometry has the intuitive appeal and simplicity which make it possible to easily obtain and establish the necessary algebraic formulas. As for literal symbolism, its use called for a level of abstraction higher than the one reached at that time.

Following Zeuthen and Neugebauer, historians of mathematics have tended to assume that geometric algebra came into being in ancient Greece, and that its rise was connected with the discovery of the problem of incommensurability. But we think that this is not so. After all, the ancient Indian *Sulvasūtras* contain no reference to incommensurable segments but do contain constructions which we have every right to classify as belonging to geometric algebra. Similarly, the ancient Chinese did not consider the irrationality of various numbers but did use geometric constructions for the solution of algebraic problems and did use geometric terminology. It seems that the use of the language of geometry was so natural that it required no special justification. This being so, it is doubtful whether the discovery of geometric algebra was connected with the discovery of incommensurable segments and whether it should be credited exclusively to the Greeks.

S. Unguru and his followers are of the opinion that Book II of the *Elements* is a purely geometric work. We regard as obvious van der Waerden's assertion that it is almost impossible to think of geometric problems that would correspond to some of its propositions. As shown earlier, the propositions in

Book II were known in ancient India and in ancient China, where at the time geometry did not exist as a theoretical science. These propositions served only as proofs of indispensable algebraic identities.

To sum up:

1. In the mathematics of all ancient civilizations magnitudes were represented by segments, and the product of two magnitudes a and b was represented by a rectangle with sides a and b.

2. In ancient Greece, India, and China there were geometric justifications of the identities

$$(a \pm b)^2 = a^2 + b^2 \pm 2ab \quad \text{and} \quad a^2 - b^2 = (a - b)(a + b),$$

and in India and Greece also of the identity

$$ab = \left(\frac{a + b}{2} \right)^2 - \left(\frac{a - b}{2} \right)^2.$$

3. There were geometric solutions of the quadratic equations

$$x^2 = ab, \quad x(a - x) = S, \quad x(a + x) = S.$$

4. There was no developed theoretical geometry based on a system of axioms either in China or in India. Thus the diagrams given earlier cannot be regarded as statements of geometric theorems (no such theorems existed there at the time) and must be viewed as records of corresponding algebraic identities or as solutions of algebraic equations.

5. It was only in ancient Greece that geometry was constructed as a theoretical science based on a system of axioms and where geometric algebra was further developed. For example, Greek mathematicians proved that multiplication is distributive over addition; investigated under what conditions the equation $x(a - x) = S$ had positive roots; provided a geometric justification for the solution of the Pell equation $x^2 - 2y^2 = \pm 1$; and so on.

Thus we regard it as obvious that geometric algebra came into being as the most convenient, natural, intuitive, and as the only possible—on a certain level of abstraction of mathematical knowledge—system for general reasoning and for the justification of known facts. In the course of their respective evolutions, all ancient systems of mathematics went through a phase like geometric algebra. And in ancient Greece geometric algebra was theoretically interpreted and acquired a completed form after the discovery of the problem of incommensurability and after geometry became a rigorous mathematical theory.

References to the Appendix

1. A. G. Barabashev, On the problem of the rise of theoretical mathematics, *Methodological Problems of the Development and Application of Mathematics*, USSR Academy of Science, Moscow, 1985, 177–187.

2. I. G. Bashmakova, Lectures on the history of mathematics in ancient Greece, *Historico–Mathematical Investigations*, issue 11, 1956, 225–400.

3. I. G. Bashmakova, The role of interpretation in the history of mathematics, *Historico–Mathematical Investigations*, issue 30, 1980, 182–194.

4. E. I. Berezkina, *The Mathematics of Ancient China*, Moscow, Nauka, 1980.

5. B. L. van der Waerden, *Science Awakening*, Noordhoff, Groningen, 1954.

6. I. O. Neugebauer, *The Exact Sciences in Antiquity*, Brown University. Press, 1957.

7. I. I. Sufiyarova, *Indian Mathematics in the Sulvasūtras and in the Works of Aryabhata and Bhaskara*. Dissertation. Moscow, 1990.

8. H. G. Zeuthen, *Geschichte der Mathematik im Altertum und Mittelalter*, Copenhagen, 1896.

9. Tsian' Bao-Tsum, *A History of Chinese Mathematics*, Peking, 1963. (Chinese)

10. Yao Fan, *Mathematical Fragments from the Treatise "Zhou be suan' zhin" and the Commentary on it by Chao Chun-Ch'ing*. Dissertation, Moscow, 1995.

11. I. G. Bashmakova, I. M. Vandoulakis, On the justification of the method of historical interpretation, *Trends in the History of Science*, 1994, 249–264.

12. H. Freudenthal, What is algebra and what has it been in history, *Archive for History of Exact Science*, vol.16, 1977, 189–200.

13. B. S. Gillon, Introduction, translation, and discussion of Chao Chun-Ch'ing's "Notes to the Diagrams of Short Legs and Long Legs and of Squares and Circles", *Historia Mathematica*, vol. 4, no. 3, 1977, 253–293.

14. J. Needham, *Science and Civilization of China*, vol. 3, 1954.

15. *The Sulvasūtras of Baudhayana, Apastamba, Katayana and Manava*. Text, English translation, and commentary by S. N. Sen and A. K. Bag, New Delhi, 1983.

16. P. Tannery, *Pour l'histoire de la science hellénique*, Paris, 1930.

17. S. Unguru, On the need to rewrite the history of Greek mathematics, *Archive for History of Exact Science*, vol.15, 1975–76, 67–114.

18. B. L. van der Waerden, Defence of a "shocking" point of view, *Archive for History of Exact Science*, vol.15, 1975–76, 199–210.

19. B. L. van der Waerden, *Geometry and Algebra in Ancient Civilizations*, Springer, 1983.

20. A. Weil, Who betrayed Euclid, *Archive for History of Exact Science*, vol. 19, 1978, 91–93.

References

1. I. G. Bashmakova, *Diophantus and Diophantine Equations*, MAA, 1997. (Tr. by A. Shenitzer)
2. I. G. Bashmakova, E. I. Slavutin, *A History of Diophantine Analysis from Diophantus to Fermat*, Nauka, 1984. (Russian)
3. N. Bourbaki, *Elements of the History of Mathematics*, Springer-Verlag, 1994. (French original published in 1969.)
4. A. Daan-Dal'mediko, J. Peiffer, *Paths and Labyrinths*.
5. F. Klein, *Lectures on the History of Mathematics in the 19th Century*.
6. A. N. Kolmogorov and A. P. Yushkevich (eds.), *A History of Mathematics from the Earliest Times until the Beginning of the 19th Century*, Nauka, vol. 3, 1972. (Russian)
7. A. N. Kolmogorov and A. P. Yushkevich (eds.), *Mathematics of the 19th Century (Mathematical Logic, Algebra, Number Theory, Probability)*, Birkhäuser, 1992. (Tr. from the Russian)
8. B. L. van der Waerden, *Science Awakening*, P. Noordhoff, 1954. (Tr. by Arnold Dresden)
9. B. L. van der Waerden, *A History of Algebra*, Springer-Verlag, 1985.
10. H. G. Zeuthen, *History of Mathematics in Antiquity and in the Middle Ages*. (German translation of the Danish original of 1893).

Additional references

1. A. D. Aleksandrov, A. N. Kolmogorov & M. A. Lavrent'ev (eds.)*Mathematics: Its Content, Method and Meaning* (3 volumes), MIT, 1963. Vol. 1, Ch. 4: B. N. Delone, Algebra: theory of algebraic equations.
2. S. Borofsky, *Elementary Theory of Equations*, Macmillan, 1950.

3. A. Clark, *Elements of Abstract Algebra*, Wadsworth, 1971.

4. C. R. Hadlock, *Field Theory and its Classical Problems*, Carus Monograph No. 19, MAA, 1978.

5. D. E. Littlewood, *The Skeleton Key of Mathematics: A Simple Account of Complex Algebraic Theories*, Hutchinson University Library, London, 1949, 1957.

6. Jean-Pierre Tignol, *Galois' Theory of Algebraic Equations*, Longman, 1988.

Index